ERN
SHACKLETON:
A LIFE OF ANTARCTIC EXPLORATION

PAUL P. SIPIERA

25 April 2019

KENDALL/HUNT PUBLISHING COMPANY
4050 Westmark Drive Dubuque, Iowa 52002

Photos on pages 99 and 110 courtesy of Corbis.

Photos on pages 61, 69, 104, 106, 111 (bottom) and 116 courtesy of Scott Polar Research Institute.

Unless noted, all other photos courtesy of Paul P. Sipiera.

ISBN 0-7872-8913-2

Printed in the United States of America
10 9 8 7 6 5 4 3 2 1

DEDICATION

To my understanding wife, Diane
and our three daughters,

Paula Frances,
Caroline Antarctica,
and
Andrea Marie

and to my fellow Antarctic 2000 Expedition members:

David G. Butts, Owen K. Garriott, William J. Gruber,
Sharon Hooper, Richard B. Hoover, James A. Lovell, Amanda Onion,
Adam Petlin, and James N. Pritzker.

CONTENTS

FOREWORD VI
By James A. Lovell

INTRODUCTION VIII

CHAPTER ONE 2
Antarctica: Earth's Final Frontier

CHAPTER TWO 14
The Learning Years

CHAPTER THREE 24
National Antarctic Expedition

CHAPTER FOUR 34
South With Scott

CHAPTER FIVE 46
Home From The Ice

CHAPTER SIX 56
Closer To The Pole

CHAPTER SEVEN 74
Yesterday's Hero

CHAPTER EIGHT 86
A Second Chance

CHAPTER NINE 96
A Long Way Home

CHAPTER TEN 118
The Last Voyage

EPILOGUE 126

BIBLIOGRAPHY 132

FOREWORD

Captain James A. Lovell, Apollo 13 commander, in the Thiel Mountains, Antarctica January 2000.

From an article in *The Wall Street Journal* by Stephanie Capparell dated Tuesday, December 19, 2000:

> Capt. James A. Lovell, Jr. knows the value of such teamwork and camaraderie. As commander of another "successful failure," as NASA dubbed Apollo 13, he also led his men to safety against incredible odds. He sees many similarities between Shackleton's ordeal and the one he faced in space in 1970 when his crew was forced to abandon its command ship for a cramped lunar module after an explosion knocked out life support systems.
>
> "*I think it's very important that everybody has a job to do and everyone chips in*," says Capt. Lovell, who read about Shackleton before a recent trip to Antarctica. "*You go in knowing everything is not going to work, and if you can think of things that can go wrong you can 'think ahead.'*"

Recently, I had the opportunity to be part of a scientific expedition to Antarctica in search of meteorites. Working out on the ice and sleeping in tents gave me a sense of camaraderie with those first explorers like Shackleton. It was an exciting moment for me when I stood at the South Pole, something that Shackleton was denied.

On these pages, Paul Sipiera has painted an exciting insight into the explorer Ernest Shackleton's trials and the disappointments he faced in trying to conquer Antarctica. Sipiera captures the spirit and determination that drove Shackleton to succeed. At the same time he points out the flaws in his character that led to his failure. On January 9, 1909 he came within 97 miles of being the first human to reach the South Pole only to be forced to turn back. I can sympathize with Shackleton. On April 13, 1970 I was within 25,000 miles of landing on the moon when an explosion forced me to turn back.

Captain James A. Lovell, Jr.
Commander, Apollo 13

INTRODUCTION

Sir Ernest Shackleton

The day is January 1, 1916. Twenty-seven men gather around "the boss" to celebrate the new year. In many ways their celebration is no different than any other that was taking place around the world that day. They wished each other good health, shared memories of family and home, and all looked forward to a better year ahead. Yet their day was very different from everyone else's for they were stranded on an icefloe adrift on the southern ocean that surrounds Antarctica.

For over a year these men have been out of touch with the rest of the world. Their adventure began with the excitement of discovering new lands. Now they have lost their ship and death is a constant companion. Their very lives depend upon the limited supplies they salvaged from the wreck of their ship. Only their will to survive and the dreams of their loved ones keep alive the hope of seeing home again.

When despair overcame these men they turned to Ernest Shackleton "the boss" for inspiration and leadership. For Shackleton, escape from this icy prison is no dream, it will happen. His only question is when? Would their rescue come in time to save the lives of all his men? This he could not know. All Shackleton could do was to reassure his men that a rescue was possible and that they had to believe in themselves if they were to survive.

What is it that drives explorers like Shackleton and his men to venture into such a dangerous place like Antarctica? For some, it is the desire to be the discoverer of new lands. Others do it for personal gain or just to satisfy their ambition. Exploration can bring glory and financial reward to the few who succeed in being the first to make an exciting discovery. History rarely remembers those who are second.

In Shackleton's time, most of the Earth's landmass had been explored. Explorers had found the long sought after source of the Nile River and expeditions had pushed far up the Amazon River Basin in search of primitive tribes of Indians. Darkest Africa had given up many of its hidden secrets and strange new cultures were being discovered on remote Pacific Islands. It was a wondrous time for exploration. Knowledge of the Earth and its people was expanding at an incredible rate. There was even fear among some explorers that there would be nothing left to discover. They were wrong!

The north and south poles presented an inviting challenge for exploration. After all, these were the most inaccessible and inhospitable places on Earth. Only the bravest explorers even thought about attempting these frozen worlds, but for those who could reach them, the rewards would be great. It was a certainty that someone would one day be the first to stand at one of the poles. This would be the person history would remember!

Ernest Shackleton was one of a restless breed of explorers. He was a man of great ambition and did not fear the unknown. He was driven by his desire to achieve great things, no matter what the cost. This did not mean that he was foolhardy, but he did take risks because that was his nature. For him there was no turning back from a difficult situation. He did not believe that success was to be had at any price. Shackleton had no desire to be a dead hero and wanted to reap the rewards that come from a successful expedition.

When it came to leadership, Ernest Shackleton always placed the safety of his men first over all else. He would never ask of his men anything that he would not do himself. If going ahead meant risking the lives of his men, he had the courage to turn back and forego the glory that may have been just a few miles ahead. Courage is not always measured by success. Sometimes it takes more courage to admit defeat and to turn back. It was this ability that made Shackleton a great leader and why his men would follow him anywhere without question, even to the frozen ends of the Earth.

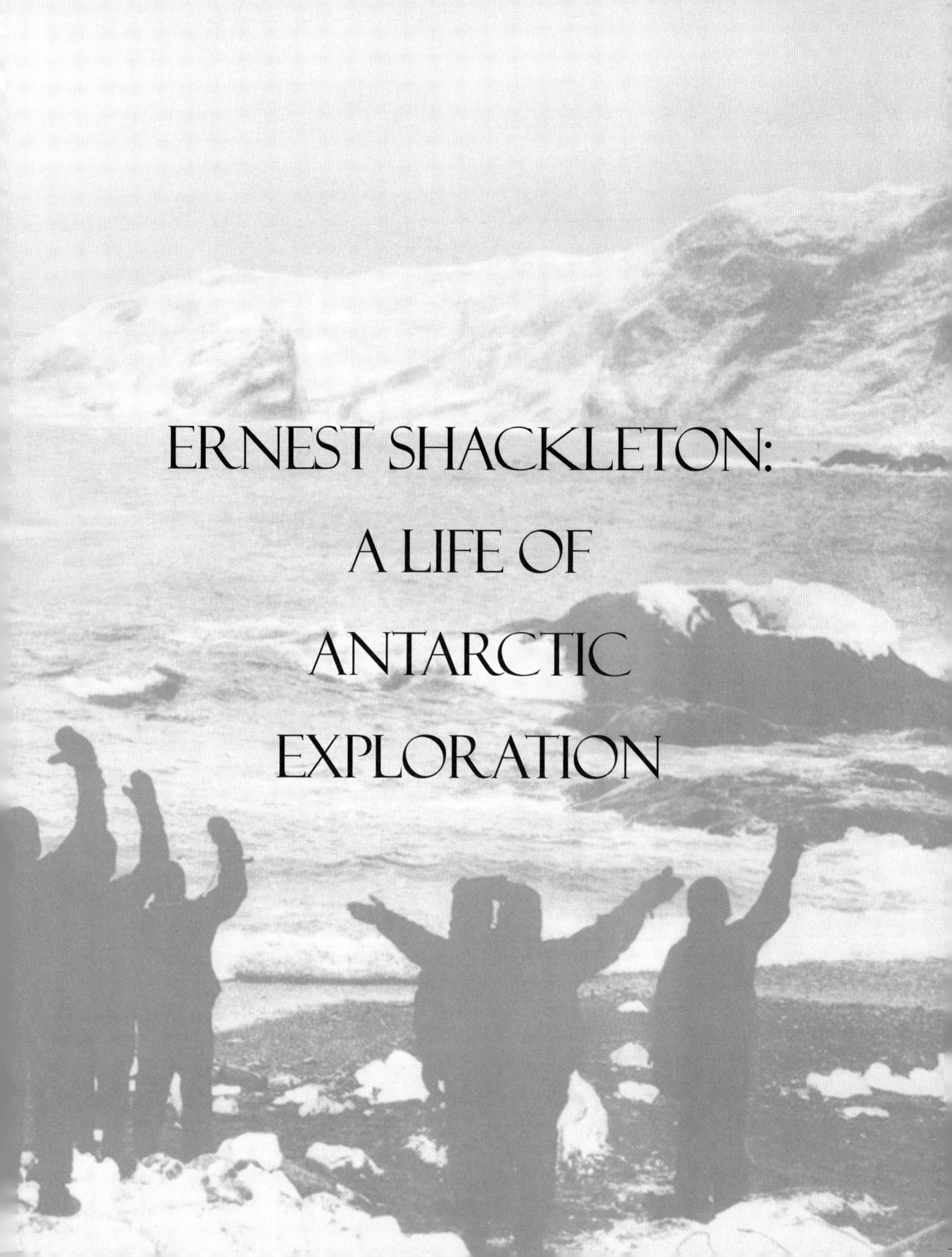

ERNEST SHACKLETON: A LIFE OF ANTARCTIC EXPLORATION

CHAPTER ONE

ANTARCTICA: EARTH'S FINAL FRONTIER

Why explore Antarctica? The dream of discovering and later exploring Antarctica is a very old one. The ancient Greeks as early as the sixth century B.C. thought that the Earth was round. By the fourth century B.C., Phoenician sailors had ventured far to the north and reported seeing huge mountains of ice floating in the sea and saw an endless land covered by snow. Later, Greek philosophers who heard of this frozen land also thought that a similar place should exist to the south. They thought, "This must be true," since the world was round. The ancient Greeks were very knowledgeable about the Earth. By the first century B.C., they had calculated the circumference of the Earth and had predicted that India could be reached by sailing west as well as east. This idea preceded Christopher Columbus by more than 1500 years. When he sailed in 1492 A.D., Columbus knew he would eventually reach India, he just did not know how far he had to sail.

In Europe, advancements in geography and mapping continued from Roman times on through the Middle Ages. People dreamed of a mysterious frozen continent to the south, but no one had ventured forth to find it. Maps that were originally drawn by Ptolemy in 150 A.D. were rediscovered and challenged explorers to seek out these unknown lands. The eventual discovery of Antarctica was only a matter of time as brave men in their tiny ships pushed further and further south. If there was a continent there, they would eventually find it.

Legend has it that around the year 650 A.D., a Polynesian sailor named Hui-te-rangiora during one of his many voyages first sighted icebergs and the ice pack that surrounds Antarctica. Early European expeditions to the Southern Hemisphere reported on the discovery of many new lands such as Terra del Fuego, the southern most tip of South America. Later, the Englishman Sir Francis Drake became the first European to sail into Antarctic waters, but it was by accident. His ship was blown far off course to the south during an attempt to sail around South America. This took him into the uncharted waters where Antarctica was predicted to be, but it was not there. Drake's venture into Antarctic waters greatly diminished hope of finding the great southern continent with all its natural riches. His reports all but destroyed interest in the mysterious southern continent, but that interest never completely died.

Dreams of finding Antarctica were rekindled through the voyages of Captain James Cook. In 1768 he was ordered to transport several astronomers to the island of Tahiti to observe a transit of the planet Venus across the face of the Sun. En route, Cook was ordered to search for evidence of the mysterious southern continent. He reported no sight of it as far as 40 degrees South. It was on his second voyage of 1772–1775 that he crossed the Antarctic Circle (latitude 66 degrees 30 minutes South) three times, but was eventually stopped by pack ice at 71 degrees 10 minutes South. His conclusion was that: "No continent was to be found in this ocean but must be so far south as to be wholly inaccessible on account of ice."

As is true with most discoveries, controversy arose as to who first sighted the Antarctic mainland. Two English sailors, Edward Bransfield and William Smith, reported sighting what they thought was the Antarctic mainland on December 20,

King penguins stand at attention at Grytvikan, South Georgia.

1819. A second sighting was made on January 29, 1820. Later studies of their ship's log proved that the December 20th sighting was actually of the South Shetland Islands and not of the mainland. A similar fate befell the January 29th sighting which proved to be Trinity Island at the tip of the Antarctic Peninsula. At the same time, an American Captain, Nathaniel Brown Palmer, made a competing claim. He claimed discovery of the mainland peninsula on November 20, 1820, and this was confirmed by a Russian expedition under the command of Captain Baron Fabian Gottlieb Von Bellingshausen. In either case, whether it to be the mainland or coastal islands, competitive Antarctic exploration had begun.

International competition for Antarctic discoveries intensified during the early part of the 19th century. Following the controversial expeditions of Bransfield and Smith, Palmer, and Bellingshausen, several nations took steps to ensure a legal claim to the great southern continent. People of this time were still convinced that Antarctica could be colonized and that vast amounts of natural resources would be found there. It was already true that enormous fortunes were being made from hunting whales and seals. If Antarctic waters possessed such riches, think what the land must hold! It was this temptation that caught the imagination of 19th century businessmen and the quest for Antarctic wealth began.

In 1830, the American Nathaniel Palmer, once again sailed for Antarctica. The principle accomplishment of his voyage was the discovery of evidence that suggested a large landmass concealed beneath the ice. Prior to his discovery, scientists believed there was such a continent, but no one had ever seen it. The first explorers only set foot on a few islands off the Antarctic coast, but did not see the actual continent. Proof came from James Eights, a naturalist on the Palmer Expedition. He observed the fact that boulders scattered about the beaches did not match the rocks which made up those islands. He correctly assumed that they were carried there by icebergs and later deposited when they grounded on the beach and melted. The source of these rocks then must come from where the icebergs had their origin, and that was the great ice sheet itself further to the south.

Deception Island, Antarctica is an active volcano with a caldera enough to permit ships to sail into it.

In 1838, the first official United States Expedition left for Antarctica. In command of a small fleet of six ships was Lieutenant Charles Wilkes. Unfortunately, none of his ships were properly equipped to sail into polar waters. Serious problems soon befell the expedition. Low morale, poor quality clothes, sickness and harsh treatment of the sailors all contributed to an unsuccessful expedition. Despite all the difficulties, the expedition did sight what is now known to be a part of the continental coastline.

Antarctic tourists enjoy the warm water heated by hot lava present just under the beach at Deception Island.

While Wilkes and his ill-prepared fleet were sailing into Antarctic waters, a British expedition under the command of Captain James Clark Ross was heading south for the purpose of collecting scientific data. One of his main objectives was to locate the South Magnetic Pole. Knowing the exact positions of the earth's magnetic poles is very important since sailors navigate by magnetic compass headings. Since Ross had already discovered the North Magnetic Pole, he felt confident that he would find the South Magnetic Pole too.

The search for the South Magnetic Pole became the first true international scientific competition. Both England and France dispatched expeditions to find it. The French Expedition was lead by Jules Sebastien Cesar Du Mont d'Urville. Curiously, the expeditions of Ross, Wilkes, and d'Urville were all exploring the same area, but they had no knowledge of the other's presence. It must be remembered that this was a long time before radio and radar. Early exploration was a very dangerous endeavor. Unless the expeditions actually passed each other, there was no way of knowing each other's position. This also meant that there was no way of helping each other in the case of an emergency.

Of the three 1840 Antarctic expeditions, the Ross Expedition was the best prepared. His ships had reinforced hulls to protect against ice and they were designed with watertight compartments to prevent sinking if the hull was breached by ice. Yet with all his preparation and excellent ships, Ross was no more successful in finding the South Magnetic Pole than d'Urville. Yet, due to the strength of his ships, Ross was able to break through the expanse of pack ice. Much to his surprise, Ross sailed south into ice-free water for almost another 400 miles. Ross thus became the first person to sail into the heartland of the Antarctic. Scott and Shackleton would later follow his route as they made their attempts to reach the South Pole.

Among the wondrous sights seen by Ross and his crew was the great volcano, Mt. Erebus, so named for one of his two ships. Much to their delight, as the two little ships sailed by, Mt. Erebus put on an impressive eruptive display. What a magnificent sight it must have been! As they sailed through what is now known as McMurdo Sound, they sighted a second big volcano. Ross named it Mt. Terror after his second ship under the command of Captain Francis M. Crozier. As the crew gazed at the shoreline, they observed a great mountain chain that extended southward, presumably all the way to the pole. Finally, they encountered the Barrier. This was an imposing cliff of ice that apparently arose out of the sea and extended south as far as the eye could see. No passage could be found through it, and with no hope of sailing further south, they turned back from the Barrier. Today the achievements of Ross and Crozier are remembered in the names of the Ross Sea, Ross Ice Shelf and Cape Crozier. It is one of the rewards of the exploration!

Mt. Erebus, a huge active volcano, dominates the landscape on Ross Island, Antarctica.

During the first 40 years of the 19th century, Antarctic exploration flourished and then stopped, only to feverishly begin again in the last decade of that century. In between, Antarctica was left to the sealers and whalers. Antarctic waters offered a rich reward for those willing to sail so far from their home. The demand for whale oil and seal fur was great and large profits made it a worthwhile adventure. The far south would be no stranger to the whalers and sealers.

The ice barrier which prevented early explorers from reaching the interior of Antarctica.

Whale bones on the beach at King George Island serve as a reminder of the whaling that still takes place in Antarctic waters.

The mysterious nature of Antarctica, as viewed by a 19th century explorer, was heightened by an interesting theory that the Earth was really hollow. An American, John Cleaves Symmes, proposed this theory in the mid-1800s. In it he envisioned a hollow Earth containing a number of concentric spheres that could be inhabited by humans. His proposed entrance to these spheres could be found at either the north or south poles. As a result of his appealing arguments and the general lack of information about the polar regions, a widespread belief in his theory became popular. It may have even inspired Jules Verne to write his novel, *Journey to the Center of the Earth*.

The classic age of Antarctic exploration began in the 1890s with the discovery of possible ancient human artifacts present on an island off the Antarctic Peninsula. On July 24, 1895, the Norwegian explorer, H. J. Bull, and an Australian schoolteacher, Carsten E. Borchgrevink, became the first people to set foot on continental Antarctica. They were crew members from the ship *Antarctic*, which was under the command of

In the nineteenth century gray fur seals like these were once hunted almost to the point of extinction.

Leonard Kristensen. Almost 75 years had passed from the first sighting of the Antarctic continent to an actual landing. The rough waters, the presence of floating ice, weak ships and the uncertainty of the weather all worked against successful landings. Later improvements in technology, better preparation and personal courage would eventually prevail and Antarctic exploration would now progress at a rapid rate.

Huge iceberg in the Scotia Sea of the coast of Antarctica.

The next year Carsten Borchgrevink led an expedition that would establish the first "winter-over" base on the Antarctic continent. During their stay they were able to climb the ice cliffs of the Ross Barrier and set foot on the ice plain that stretches all the way to the South Pole. Later expeditions led by Shackleton and Scott would push the extent of "further-south" to within 97 miles of the South Pole. Eventually, in 1911, it would be Amundsen that would reach the South Pole first. Clearly Antarctic exploration had evolved from whalers plowing through its waters seeking wealth, to scientific exploration whose goals were discovery and the accumulation of new knowledge.

As the 20th century began, the question of a large landmass beneath the ice was still unresolved. Was the land sighted by the first explorers actually a continent, or was it just part of a series of large islands? A Canadian scientist, Dr. John Murray, summarized all available evidence in February 1899 and supported the theory of a large lost (buried under the ice) continent. He based this on a few rock specimens brought back by several of the earlier expeditions. These rocks were very supportive of a continental landmass. Knowing the potential mineral wealth that a continent can hold served as a great incentive to pursue Antarctic exploration. It now became the beginning of a race between nations to see who could secure the most valuable territory.

Although the romantic challenge of being the first to stand at the South Pole was publicly appealing the real reason behind exploration was in discovering new natural resources. The industrial revolution was now over a hundred years old and the known resources were being consumed at an alarming rate. New resources had to be found to meet the ever-increasing demands. The idea of a new untouched continent was irresistible for both industry and governments alike. This was the kind of world in which the likes of Ernest Shackleton, Robert Falcon Scott and Roald Amundsen lived. The lure of great wealth and the spirit of national pride led these men forward on the path of exploration. The opportunity for fame and fortune was there for the taking and they answered the call. Some met with success and others met with defeat, but in the end, each achieved their goal. They pushed back the frontiers of discovery and the world remembers them for it.

CHAPTER TWO

THE LEARNING YEARS

THE LEARNING YEARS

The continent of Antarctica provides the occasional visitor with peace and solitude that very few other places on Earth can offer. Crisp, clear air highlights this vast ice desert landscape with a sky so large that one can lose sight of reality and a sense of time. This is Antarctica on a clear day. When the storms rage, the same continent is capable of stirring up the most ferocious weather on Earth. For Antarctic explorers, time on the ice can pass both very quickly or move as if time had stood still. It depends upon the situation. If the weather is fine and there's work to be done, time flies by. As the fierce wind swirls the snow so hard that it creates a white blindness, there is little else to do but huddle in your tent and try to stay warm. Snowstorms could last for days, and the explorers had little to do but think and reflect upon what brought them to Antarctica. What drives a person to push himself to the ends of the Earth? Shackleton may have asked himself that question many times during his four expeditions to Antarctica. Although there is no one answer to such a question, from his earliest years he had the spirit of an explorer. Ernest Shackleton may have been destined to become an explorer, but no one would have guessed that from his childhood.

Ernest Henry Shackleton was born in the small town of Kilkea, County Kildare, Ireland on February 15, 1874. He was the second child born to Henry and Henrietta Shackleton. Their family would be a large one, eventually totaling two boys and eight girls. The Shackletons were English-Irish gentry who made their living by farming the land. They lived in a large house that was built in the late 18th century in the Georgian style that was popular at that time. The town of Kilkea was surrounded by rich farmland, yet it was located only 30 miles away from the society life of Dublin. Young Ernest would learn from both environments. He would grow up with the values of a simple, hardworking farm life, but also with the ambition and worldliness that the city had to offer.

The Shackleton family descended from ancestors that lived in Yorkshire, England, in the early part of the 18th century, a time when all of Ireland was an English colony. The Shackletons were part of a movement to settle Protestant English in Ireland in an attempt to develop a ruling class over the native Irish who were Catholic. A large immigration of settlers would play an important role in the English plan to rule Ireland. For the most part, the plan failed. Irish resistance was strong enough to break any yoke the English tried to place on it. Many of the early English settlers also became so much a part of their new homeland that they eventually became true Irish, resisting English rule alongside their native brothers.

At the time of Ernest Shackleton's birth, Ireland was a troubled land. The majority of the Irish people wanted Home Rule, a form of self-government from England. The English, on the other hand, were determined to maintain their hold on the Irish and violence erupted many times. This created an unrest among the Irish people and divided their loyalty between their native land and those who ruled them. If that was not enough in itself, a crop failure occurred and that completely changed the Irish way of life. The Irish economy depended heavily on its potato crop and when their harvest was repeatedly low, poverty swept the country. Life for the Shackletons would be different if they were to remain in Ireland. Something had to be done if the family were to have a comfortable life.

In the face of economic depression and political unrest, Henry Shackleton made the difficult decision to leave the farm and move to the city. Not only would the Shackletons leave Kilkea behind, but Henry would begin an entirely new career as well. These were indeed difficult times for the Shackletons as they embarked for Dublin. Then in 1880, Henry Shackleton enrolled at Trinity College to begin his studies as a doctor of medicine. This entire change of events must have made a big impression on young Ernest. First they left behind the gentle country life, and now they were moving into the fast-paced and exciting life of the city. It was a big change for all.

Upon completion of his medical studies, Henry Shackleton made another big decision that would affect the entire family. They were to move again, but this time to England where he hoped their life would be much better. Henry felt that the English-Irish situation would only worsen and the safest place for his family would be England. As a result, Dr. Henry Shackleton opened a medical practice at Sydenham on the outskirts of London. It was not long before Henry proved himself right as his medical practice flourished and he was able to properly provide for his family. It would be here at Sydenham that Ernest would grow into manhood.

Ernest Shackleton's mother was a woman devoted to her home and family. She maintained a secure and happy home and expressed a certain optimism that carried her family through the difficult times. It was very unfortunate that she took ill shortly after their arrival in England. She became an invalid and no longer took an active role in family life. This left the raising of the children up to Henry who enlisted the help of several female relatives. As a result, the Shackleton household was dominated by women. Ernest and his brother were certainly outnumbered by his sisters and the parade of women relatives that appeared to take care of them. Ernest quickly learned to cope with the situation and through charm, turned it into an advantage for himself. He was often accused of playing one sister against another. He also learned that an expressive face can change a punishment into a reward if done just right.

As a child, Ernest was really no different from any other. He got into the usual trouble that most boys do, and was looked upon as a leader even at an early age. He had no difficulty in persuading people to his way of thinking. Ernest Shackleton's early education was conducted at home by a governess and later at Fir Lodge Preparatory School. His fellow students remembered him as being friendly and good natured, but also one who would be in the middle of every fight. His Irish accent often caused him trouble when the English boys made fun of it, and Ernest was not about to let them get away with it. This gave him one of his many nicknames "the fighting Shackleton."

At age 13, Ernest left Fir Lodge Preparatory School for Dulwich College. There, for the most part, he was a loner. He would prefer a good book and his own company to that of joining into group activities. He did not participate in the popular team sports that most boys played in and he remained alone as often as possible. He also didn't like the traditional subjects of Latin and Greek that the school taught. He preferred the more useful modern languages and mathematics. To sum it up, school was a big bore for Ernest Shackleton and he felt very much out of place and dreamed of an exciting future.

One thing that young Ernest did enjoy at school was reading a weekly publication called *The Boys Own Paper.* In it were stories about science fiction, world travel, and especially dangerous adventures on the open seas. He would read these with great interest and dream about high adventure and distant places, with himself as the hero in every story. It certainly must have made a lasting impression on the boy who would grow up to explore the far reaches of the Earth.

At school, Ernest Shackleton displayed a certain disregard for rules and was often late for classes. His colorful excuses for being tardy became very entertaining for his classmates. The Irish have a reputation for storytelling, and Ernest fulfilled his role. Perhaps this certain creativity impressed his teachers enough to forget about any serious punishment. This kind of behavior was not what his father expected from him. His father had hoped that Ernest would follow in his footsteps and become a doctor, but this was not to be. Ernest constantly talked about going to sea and finding a career either with the navy or in merchant shipping. In his day a boy could apply for sea training at the age of $14^1/_2$. His duties at sea would be hard, but it would train him to learn discipline and a useful trade in which he could eventually earn a good wage.

Shackleton's desire to go to sea was a little ironic, since he could not swim. It was during a seaside holiday with some school friends that Ernest almost lost his life. He was playing in the water with some other boys and because Ernest could not swim properly, he tired quickly. Ernest was in trouble and began to drown. Had it not been for his best friend pulling him in to shore, Ernest would have drowned. Shackleton would never forget his friend, Nicetas Petrides, the boy who saved his life.

In time, Henry Shackleton realized that he could not keep his son in school so he reluctantly consented to let Ernest join the merchant marine for sea training. His first ship would be the *Hoghton Tower*, a square-rigged sailing ship. Shackleton's first voyage would take him around Cape Horn, the southern tip of South America, where the seas are said to be the roughest in the world. It would be a good test for the lad of 16 who wanted to go to sea.

It took excellent seamanship to properly sail a square-rigged ship. Shackleton had to learn the names and uses of over 200 different ropes and how to tie many different kinds of knots. This was very different from what he was learning at school. Failure to recognize a particular rope during a storm might endanger the ship, and men's lives could be lost as a result. Making a mistake on board ship resulted in a punishment that Shackleton could not talk his way out of. Wrong answers met with the sting of a rope across his back or hands and he learned very quickly not to make mistakes.

The *Hoghton Tower's* destination was Valparaiso, Chile. To get there the ship had to fight its way through high seas and strong winds for nearly two months before it made it around Cape Horn. For days on end the high winds would push against the ship and it appeared that no forward progress was being made. To sail into the wind, a ship has to perform a tacking motion (zig-zag course) against the prevailing westerlies. It was very rough going and it pushed the ship and crew to their limits of endurance. In addition, they were close enough to Antarctica to be worried about icebergs. A constant watch had to be kept in order to avoid a collision with these floating mountains of ice. In his first sea voyage, Ernest realized all the adventure he had ever dreamed about while reading his books about life at sea. The earlier excitement soon turned into boredom and his dreams were now focused on reaching the next exotic port.

Upon returning home from his first voyage, Shackleton was faced with his first real-life decision. Would he return to sea and seek a career as an officer or would he go back to school and follow his father's wishes? The sea would be his choice and he returned to the *Hoghton Tower*. After making several voyages to South America and the Orient, Shackleton took his officer's examination and passed it in August 1894.

Life at sea had a very big influence on Shackleton's life. Just as he did in school, Ernest spent a good deal of his free time on board the ship by himself. He was constantly reading and he especially enjoyed poetry. Often, while he was on watch, he could be heard reciting poetry as the ship rose and fell with each passing wave. He was acquiring the reputation of being a scholar. This was very different from his experiences at school!

After spending four years at sea, Shackleton came to realize that the future in shipping was with steam and not sail. He decided to further his career by qualifying for a First Mate's Certificate. His first berth on a steamship was as Fourth Mate on the *Monmouthshire*. He would spend the next five years sailing to the Orient and the Americas. During those years he spent more time reading and studying. He was preparing for the examinations that would advance him in rank. Shackleton achieved his First Mate's Certificate in 1896 and was later certified as a Master in April 1898. He was just 24 years old and now he was qualified to command a British ship anywhere in the world. It was a tremendous accomplishment for a man who eight years before had only dreams.

Ernest Shackleton's early years at sea were not entirely spent in seclusion. He enjoyed his time at home with his family very much, especially with his sisters. They had wonderful times together when he was home from the sea. It was during one of his shore leaves that he met Emily Dorman, his future wife. Emily was a friend of his sister, Kathleen, and she quickly caught Ernest's eye. They spent as much time together as possible and shared in their mutual love of Browning's poetry, but the fate of every sailor kept them apart.

Shackleton's prolonged periods at sea were made easier by exchanging many romantic letters with Emily. He missed Emily very much while he was away and it even made him think about leaving the merchant marine to be with her. Finally, he decided to give up the long cargo voyages for the shorter passenger lines. The advantage in doing this was in their regular schedules and predictable destinations. This would make it much easier to see Emily more often.

Clearly, marriage to Emily was on Shackleton's mind. Ernest was also very aware that Emily's father was not happy with the prospect of her marrying a common seaman. He felt that if Emily married Shackleton, she would be marrying beneath her social position. Shackleton, on the other hand, was very proud of his family heritage and felt he was worthy of Emily. Ernest was still sensitive to her father's concern and sought ways to prove him wrong. To this end, promotion in rank and having the opportunity to make a name for himself were very important. This is one reason why Shackleton did indeed leave the merchant marine for passenger service. Service with the Union Castle Line would give him the opportunity to meet many influential people. Shackleton was a true opportunist who was always looking for ways to better his position in life. His ambition would not stop here, but would eventually take him literally to the ends of the Earth.

CHAPTER THREE

NATIONAL ANTARCTIC EXPEDITION

As the 19th century drew to a close, explorers from around the world saw Antarctica as a competitive arena and the prize would be fame and fortune. At this time, the British government looked to Antarctica as both a place of scientific discovery and as an opportunity to test the ability of its naval officers. The British Navy hoped that Antarctic exploration would test the ability of its sailors and help them develop war time skills. Maintaining a high level of proficiency in a peace time navy is very difficult. Service in Antarctica appeared to be the solution. Its seas are among the roughest in the world and navigating safely through pack ice and icebergs requires good judgement and great skill. To this end, the British Navy encouraged its officers to participate in Antarctic exploration. Among the first to volunteer for Antarctic service was Robert Falcon Scott.

The British National Antarctic Expedition of 1901–1904 was the dream of Sir Clements Markham, the president of the Royal Geographic Society. For many years he had been promoting the idea of a national Antarctic expedition and had finally convinced the Navy and other scientific societies of its merit. The man he chose to lead the expedition was a young naval lieutenant named Robert Falcon Scott. Sir Clements had met Scott years before when Scott was only a naval cadet. That first meeting with Scott greatly impressed Sir Clements, and years later, he still remembered him. Markham worked very hard to have Scott assigned as the leader of the expedition. In one sense, this was rather odd. Scott did not possess the naval background nor Antarctic experience necessary to lead such a demanding expedition. Sir Clements must have seen something in Scott, and it must have been his obsession to succeed. After his appointment, Scott enthusiastically set out to learn all he could about polar exploration. He knew that his future in the Navy was tied to the success of this expedition. Success in Antarctica would bring fame and even a fortune too. This is exactly what Scott had hoped for.

At the time when the National Antarctic Expedition was being organized, Ernest Shackleton was serving as an officer on one of the Union Castle Lines' passenger ships. Although he held a prestigious position, Shackleton desired more. He felt that if he could become involved in some well-publicized adventure, it would draw attention to his abilities and he could make a name for himself. A notice placed in the daily newspaper caught his eye. It called for volunteers to participate in an Antarctic expedition. On September 13, 1900, he wrote to apply for a position, and four days later, presented his case in person at the expedition office in London. Although Shackleton was an able and experienced sailor, luck and influential friends played an important part in his being chosen for the expedition.

The *Discovery* Expedition, as it would later be called, had to face some difficult problems right from the start. Sir Clements Markham had originally hoped that the expedition would be conducted by the Royal Navy. This was not to happen. The British government was reluctant to assign its naval personnel to any but military assignments. All available military personnel were needed in South Africa to participate in the Boer War. Sir Clements finally had to settle for a publicly-funded expedition staffed by civilian sailors. He did succeed in receiving the loan of a few

Royal Navy officers to command the expedition. Sir Clements was confident that having Royal Navy participation would better his chances for a successful expedition.

The combination of Royal Navy officers in charge of civilian sailors led to serious conflicts between the two groups. The Navy personnel were used to strict discipline and doing things the military way. The civilian sailors in turn felt as if they were being treated unfairly. Emotions ran high and bad feelings soon developed between the two groups. This led to many uncomfortable situations and rivalries soon developed. These circumstances demanded a very strong leader who could balance each side and blend them into a strong team. Unfortunately, Scott was not the answer to the problem. He usually refused to become involved in personnel matters and let his subordinate officers solve the problems.

As the *Discovery* Expedition slowly began to come together, the situation between officers and crew became so serious that consideration was given to replacing Scott as the expedition's leader. In an attempt to save the expedition and Scott, Sir Clements Markham assumed the role of arbitrator. He kept both sides focused on the expedition and diverted their attention away from any conflicts and personal problems. It was a delicate situation, and Sir Clements succeeded in doing what Scott avoided.

Ernest Shackleton was at sea when word came of his acceptance to the *Discovery* Expedition. When he heard the news, he was delighted, and anxiously waited for a meeting with Scott. No one could have imagined that this first meeting would be the creation of two legendary Antarctic heroes. Each man possessed similar ambitions and comparable sea experience, and neither had any Antarctic experience. In fact, before the announcement of the *Discovery* Expedition, they never even considered Antarctica as their road to fame and fortune. Fate, it seems, may have taken a hand in bringing them together. Not as cooperative partners working toward a common goal, but as rivals fearful and jealous of each other. Neither Scott nor Shackleton intended it to be that way, but as the expedition progressed, their competitive nature simply took over. Circumstances later proved that Shackleton possessed a stronger personality and was the natural leader. Scott, on the other hand, distanced himself from his men and kept to himself. Their contrasting personalities were noticed by the other expedition members, and many of the men began to favor Shackleton over Scott as leader.

Now as a member of the *Discovery* Expedition, Shackleton had to take leave from his duties with the Union Castle Line. His first role was to work in the expedition's headquarters in London. The months of preparation required for the expedition suited Shackleton very nicely. There was much to do and he found himself right in the middle of things. He had to attend fund raising dinners and other public events. This gave him the opportunity to meet many influential people. Living and working in London also gave Shackleton time to court Emily Dorman. If he was ever to marry Emily, he would first have to convince her father that he was worthy of her. Moving about in London's social circle certainly did not hurt his chances of impressing Mr. Dorman.

Discovery was the name of the ship that would eventually take Scott and Shackleton to Antarctica. She was especially built for the expedition at the famous shipyards at Dundee in Scotland. *Discovery* was a three-masted, square-rigged barque that was designed to use either sail or steam. In open sea she rode the waves with a distinct rough lurching motion. Also, as is typical of all wooden ships she leaked, but considerably more than most. This was a serious problem and pumps had to be used all the time to kept water out. It was part of Shackleton's duties to maintain a watch in the cargo hold on this

A model of Scott's ship Discovery *as seen in an exhibit at the Canterbury Museum in Christchurch, New Zealand.*

leaking problem. Another fault in the ship's design was not enough room to store a sufficient amount of coal. This severely limited the ship's operational range when under steam power.

Scott was very impressed with Shackleton's extensive experience at sea and chose him to oversee part of the ship's construction. In addition, he was directed to supervise her sea trials. It was Shackleton's duty to evaluate her performance. He quickly learned how the ship behaved under different conditions, and how to compensate for her faults. Scott did not participate in this part of the expedition since he was involved with other organizational matters and fund raising. Selection of vital equipment and the ship itself were left to subordinates like Shackleton. Scott placed a great deal of confidence in the ability of these men, and Shackleton was no exception.

Preparations for the departure of the *Discovery* Expedition experienced a series of ups and downs. At times, the lack of funds and political support seemed to doom the expedition. Finally, after all preparations were made, *Discovery* received an inspection visit from King Edward VII and Queen Alexandra. The excitement of the royal visit gave the Antarctic-bound expedition the morale boost it needed. Shackleton saw the royal visit as the ideal opportunity to write Mr. Dorman for permission to marry his daughter, Emily. He felt that Mr. Dorman could not refuse him since he was part of an expedition supported by the King of England. Shackleton left England with a very good feeling about his future. He was certain that he would make a name for himself and, upon his triumphant return, he would marry Emily. Shackleton's star was certainly on the rise.

Finally, with all preparations completed, on Tuesday, August 6, 1901, *Discovery* set sail for Antarctica. *Discovery's* 50-man crew was eager to reach their destination. Shackleton was one of Scott's deck officers. Also included among the ship's crew was a young surgeon and zoologist named Edward Wilson. It would be the trio of Shackleton, Wilson, and Scott who would make the first attempt to reach the South Pole and their names would be forever linked to Antarctic exploration.

The route Scott chose for *Discovery* was one very familiar to Shackleton. Scott's plan was to sail around Cape Horn at the tip of South America and then on to New Zealand. Several years before, during his time with the merchant marine, he had sailed this route many times. The voyage to New Zealand would prove to be the first extended test at sea for *Discovery*. Once in New Zealand, the expedition would rest and make final preparations for the final leg to Antarctica.

Discovery's original design was to be a polar exploration vessel, but was later drastically changed by Sir Clements Markham. The final design limited the effectiveness of the ship in polar waters. The leak that was first discovered during her sea trials became worse and remained a serious problem throughout the expedition. In addition, *Discovery's* deck was seriously over-loaded with extra coal, sheep, and sled dogs. It also held more scientific equipment, including a hot air balloon, than any other previous expedition. All of these factors made *Discovery* a very difficult ship to sail. Since Shackleton had the most sea experience, he took it upon himself to be personally responsible for his ship's safety.

In spite of the concern over the many problems, *Discovery* set sail from Lyttelton Harbor, New Zealand on Christmas Eve, 1901. Luck was on *Discovery's* side as she sailed the 1,100 miles without serious trouble. The roughest seas in the world favored *Discovery* for the moment, and on January 2, 1902 she reached the pack ice that surrounds Antarctica.

The Royal Hotel in Lyttleton, New Zealand served explorers en route to Antarctica.

One week later, *Discovery* made her way through the pack ice and reached Cape Adare. A sudden shift of the tide swept the ship out toward a group of icebergs and certain destruction. Luck once again favored Scott as the tide shifted a second time and *Discovery* was able to make for open water. This narrow miss with disaster demonstrated to everyone the danger and unpredictability of Antarctica. Nothing could be taken for granted. It was very unfortunate for Scott that he seemed to have forgotten this first lesson. Ten years later, the unpredictability of Antarctica would cost him his life, along with that of Wilson and three others.

Discovery was not the only ship in the Antarctic at this time. A Swedish expedition led by Otto Nordenskjold was on the opposite side of the continent headed for Grahamland. His ship *Antarctic* became frozen into the pack ice of the Weddell Sea and could not escape. It was eventually crushed by the ice and sank. The rescue of its crew was a long time in coming and, in order to survive, they had to endure great hardships. The sinking of *Antarctic* was very significant for Shackleton because it would later serve as an example of how to survive, when his ship *Endurance* would be lost to the pack ice too.

Cape Adare was the landing site for the *Discovery* Expedition. This was the place where the explorer Borchgrevink and his men spent the first winter on the Antarctic continent. After leaving Cape Adare, Scott sailed southward far into the Ross Sea. As they sailed along the coast, Shackleton was overwhelmed by the sights he saw. Huge mountains rose straight up out of the sea, and enormous icebergs drifted by the ship. Occasionally they could see the spouting of whales and watch penguins darting in front of the ship as they sliced through the icy water. Overhead, the ever present sun created a world beyond belief for these first-time visitors.

The first goal of the *Discovery* Expedition was to survey the extent of the great ice barrier that was first noted by Captain Ross in 1840. This huge wall of ice extends for over 400 miles and at times towers as much as 1,000 feet high. This is where the great ice floes from the interior of the continent eventually met the sea. It appeared to be an impenetrable barrier for exploration into the interior. If the South Pole was to be reached, Scott had to find a way over it.

Borchgrevink and his expedition landed near this barrier of ice and proved that explorers could travel over it. Yet, they could not determine the extent of the ice. Scott had an interesting idea, he would launch the first balloon flight in Antarctica. From an altitude of 600 feet, Scott became the first man to survey the ice barrier from above. Shackleton would follow in the next ascent, and he took photographs to record what they both had seen. The view from the balloon revealed a vast expanse of ice leading off to the south, as well as a natural harbor that would serve as their base of operations. The way to the south was now clearly open.

Home for the *Discovery* Expedition was located in a small bay at the head of McMurdo Sound. Its geographical position was determined to be 77 degrees 50 minutes South latitude. Here was an ideal location for Scott to conduct his training and scientific surveys. It also proved to be an excellent staging base for both of Scott's expeditions, and later for Shackleton's attempt at the South Pole. Today it still serves as the "gateway to Antarctica" and is principal base for the United States Antarctic Program.

A modern day view of McMurdo Station, "Antarctica's largest city," with Scott's Discovery *Hut in the foreground.*

CHAPTER FOUR

SOUTH WITH SCOTT

At the time of the *Discovery* Expedition, very few people had any experience in polar exploration. Those who dared to reach the poles would first train in Greenland, Norway or Sweden. As they gained more polar experience, they invented better equipment and learned more efficient ways to travel in arctic regions. Scott was no exception. He read as much as possible about polar exploration and interviewed those who had been there. It is unfortunate that Scott did not listen to the voice of experience. Instead, Scott drew his own conclusions about what polar exploration would be like. When he should have chosen dog sleds and skis, he went ahead and purchased expensive and unproven equipment which quickly failed in Antarctica. In this respect, Shackleton and Scott were very much alike. In their passion to succeed they both failed to follow the fundamental rules of exploration. In the end, their reluctance to listen to good advice cost Scott his life, and Shackleton the South Pole.

Antarctic explorers still use the basic tent design used by Robert Falcon Scott in 1901.

From the start, the *Discovery* Expedition ran into trouble. The inexperience of the men quickly became apparent as they attempted to drive dog sleds, set up tents and even light stoves. The simplest tasks became almost insurmountable. Everything they eventually accomplished was done with great difficulty. They even had the wrong clothing for the work to be done. Although their clothing was waterproof, it was tight-fitting and trapped both heat and moisture. The result was that sweat built up and soaked their inner clothes. This quickly became uncomfortable and often led to frostbite. A man could not work efficiently under these conditions, and his time outdoors was severely limited. In a land where food, water, heating fuel, and shelter all must be brought in, time is always working against the explorer. When the supplies run out, one dies very quickly in Antarctica.

Once the men settled into Antarctic life, Scott planned on taking several excursions away from base camp. These trials were designed to test their equipment and the men who would use it. Ernest Shackleton had the honor of leading the first excursion. These ventures beyond camp were not very productive and little was learned from them. In one particular case, a seaman was killed as he fell over an ice cliff. What these trials did prove was that travel in Antarctica would be hazardous and extremely difficult. Shackleton learned quickly and enjoyed his experiences with the wondrous beauty of Antarctica. It instilled in him the passion for discovery. He also learned that he preferred to be a leader rather than a follower. This set the stage for an irreconcilable confrontation with Scott that resulted in Shackleton's early return home.

Recreation of an early Antarctic explorer's hut at the Canterbury Museum in Christchurch, New Zealand.

Living for long periods of time with the isolation of Antarctica can affect people in different ways. Some enjoy the solitude and keep to themselves, while others go mad from boredom. Minor disagreements over simple matters can easily get out of control and escalate into a serious problem. It is not certain when the relationship between Scott and Shackleton first became strained, but their different personalities would eventually bring them into conflict. The way a leader handles his men under these conditions is very important to an expedition. Scott chose to conduct his expedition according to the strict naval tradition which separated officers from the ordinary seaman. This created a division between the two groups and the men were instructed to refrain from associating with the other. Even though Shackleton was an officer, he frequently mixed with the seamen. This attitude was very different from Scott's approach to command. It had the effect of undermining Scott's decisions and made the crew more comfortable with and trusting in Shackleton. Scott began to feel that the men looked more toward Shackleton as their leader, and he could not let that happen.

The main goal of the *Discovery* Expedition was to get as close to the South Pole as possible. Reaching the actual pole would have been wonderful, but their earlier attempts at polar travel made it unlikely even to Scott. Shackleton wanted to be a part of Scott's polar party, but he didn't think his chances for selection were very good. There were several others who were better qualified and had more experience. It was a big surprise to Shackleton when Scott chose him. Edward Wilson, the doctor, joined them as the last member of the three-man party that would proceed on to the South Pole. How far would they get? No one really knew, and could not even make a good guess. It all depended upon good weather and a bit of good luck.

Preparations for the southern journey, as the attempt for the South Pole was called, began in early September 1902. These preparations included several training excursions to see how well the men and their dog teams could work together. The results were not encouraging, since the men were not experienced dog handlers. Only a few had any experience driving dogs, and they too had to learn through trial and error. Trials like these clearly demonstrated the weaknesses of the expedition and did not build upon its strengths. Regardless of all the difficulties encountered, the southern journey departed on November 2, 1902. Amazingly, it would not be poor training that stopped them from reaching the South Pole, but the dreaded disease—scurvy.

Seals shot for food in 1910 are still frozen and ready to be eaten at Discovery *Hut.*

Scurvy is a condition that is caused by lack of vitamin C. In the days of early exploration, scientists did not know much about vitamins. Scurvy had always been a common occurrence on long sea voyages, and its cure was fresh food like meat and fruit. Scurvy mainly attacked sailors because they usually ate canned food that was devoid of vitamin C. Antarctic expeditions were modeled after long sea voyages and carried the same foods as the Navy. This was a mistake. Earlier Antarctic expeditions had found that eating seal meat prevented scurvy. Scott would not hear of it. He felt it was morally wrong to kill seals for food when tinned meat was already available. Shackleton bitterly disagreed with Scott and the suggestion of eating seal meat led to their first serious confrontation. It took a major outbreak of scurvy to change Scott's opinion. Scott was away from camp when the outbreak occurred and seal meat was served to the stricken men. All were quickly cured by eating the fresh meat.

Scott, Wilson and Shackleton departed in a festive fashion on the southern journey. Cameras were clicking off, flags were flying and cheers were heard all around camp. The inexperience of the group clearly showed as they moved away. Their five sleds were all hooked together and pulled by one large team of 19 dogs. This was the most inefficient manner of sledding possible, but no one in the expedition knew any better. This, combined with the fact that the three men refused to use skis, greatly slowed their progress. Yet, food would still prove to be the ultimate problem. The main diet for a polar explorer was based around pemmican, a dried meat mixed with fat. This was supplemented with cocoa, chocolate, biscuits, sugar, oatmeal and small amounts of cheese. Fresh meat or fruit were not considered necessary. Although the men had eaten small amounts of fresh seal meat before leaving, they were still not receiving enough vitamin C in their diet. Scurvy would soon attack them and work against the success of the party. Ernest Shackleton was the first to feel the effects of scurvy. At the time of their departure, Shackleton was still not fully recovered from his first bout with scurvy, and his already weakened condition would only get worse.

Scott's men left behind a large quantity of supplies in Discovery *Hut. A half-eaten seal stew still remains on the stove.*

Plodding slowly over slippery ice and pushing through drifting snow, the three explorers pressed on into unexplored territory. Progress was slow and the dog teams became more of a hindrance than a help. The dogs were controlling their drivers and the men did not know how to handle them. Finally, it came down to the men hauling some of the sleds themselves. This really slowed their progress. It took only a couple of weeks of this to convince them that they couldn't make it to the South Pole. The stress of man-hauling their sleds definitely brought out the personality differences in Scott and Shackleton. Once the reality of not reaching the South Pole set in, Scott became easily annoyed with his companions, and in particular, Shackleton. Any mistake made by Shackleton was met with harsh words from Scott. Dr. Wilson found himself constantly trying to keep peace between Scott and Shackleton. And if things couldn't get worse, they did, as their food began to run out. Earlier Scott allowed the men to eat as much as they wanted, now it was apparent that they did not have enough food to make it back. They would have to ration their supplies very carefully to survive.

The dogs did not fare any better than their masters. Without proper handling, the dogs were constantly being overworked. Their food supply had rotted and eating it made them sick. One by one, the dogs began to die. In the end, all of the dogs had to be killed. This experience left both Scott and Shackleton with a bad opinion of dog sledging. It was an unfortunate mistake that would later cost both of them the honor of being the first to reach the South Pole. Roald Amundsen, the man who would be first to the Pole, relied on dog sledging and probably would not have made it without his dogs.

Seven weeks out from his base, Scott found his party at 81 degrees South. They had covered just over 3 degrees of latitude, and at that pace they would never make it to the South Pole. At that point, it was questionable just how far they could actually get. Certainly a respectable record for furthest south was not going to happen, but they continued on despite their disappointment and weakened condition.

The end of the southern journey came on December 30, 1902. Exhaustion, hunger and the first serious signs of scurvy brought the expedition to a halt. The position they recorded in their logbook was latitude 82 degrees 17 minutes South. This was a new record for distance south, but it was rather unimpressive in comparison to the latitude 86 degrees 14 minutes North that the Norweigan explorer Nansen had accomplished in 1890.

The journey back to base at McMurdo Sound would not be easy. On the outward part of their trek, Scott left behind several depots of food for their return trip. This made good sense since they could travel faster by carrying a lighter load. The only danger in leaving food behind was in finding it on the way back. Blowing snow, white outs and a general loss of direction made finding the depots very difficult. Without these supplies, Scott's party would certainly die.

Time was working against Scott and the others as the austral summer was quickly fading and the weather was turning bad. On the southward journey, the weather had been extremely kind to Scott and this gave him a false sense of security. Now blizzards and white outs were much more common and the fierce winds cut through their exhausted bodies. Weakened as they were, the men could not keep the same pace. Without the dogs, and their inability to ski properly, they could only cover short distances each day. All this, combined with constant hunger pains, brought on severe depression and they began to wonder if they would make it back alive.

By mid-January, Shackleton's health took a turn for the worse. He was experiencing loss of breath and a constant cough. Despite all of his discomfort and illness, he kept pushing on with his companions. On January 18 Shackleton eventually collapsed from chest pains and the desperate march came to a halt. They were losing precious time, but there was no other choice. Shackleton could not be carried, so they had to give him the opportunity to rest. Prior to Shackleton's collapse, the men were plodding through the snow with their skis carried on the sled behind them. In an act of desperation, they asked Shackleton to put on the skis to see if they would help keep him from falling. Much to everyone's surprise, he was able to travel much easier and kept up with them much better.

One of the side effects of scurvy is the way it affects a person's mind. It makes one's judgement seem irrational at times, and minor irritations become blown out of proportion. This is how it affected Scott. He reacted badly to Shackleton's weakened condition and constantly blamed him for their slow progress and failure to make it to the South Pole. This situation worsened until Dr. Wilson had to step in and tell Scott to stop harassing Shackleton. Then things changed as Shackleton's condition temporarily improved. He was permitted to ski ahead and scout for the landmarks that would lead to their food depots. But that was not to last for long and the worst was yet to come.

As they reached the last food depot, Shackleton fell extremely ill. They were still 60 miles from their base at McMurdo Sound and Shackleton simply could not move. Dr. Wilson did not expect him to live through the night, but he did. The next day it was decided that they must reach base as soon as possible if Shackleton was to live. By shear will power and the desire to survive, Shackleton got up and moved on with the others. He would walk or ski as best he could, and somehow he made it back with the others. On February 3rd three worn out Antarctic travelers made it back to the warm cheers of their comrades. For Shackleton, he was just glad to be alive.

Both Scott and Shackleton failed in their first attempt at glory. For Scott, he would remain in Antarctica for another year creating a successful expedition. It would be different for Shackleton. Scott ordered Shackleton, with seven other men, to return home on the relief ship. This was not the glorious homecoming that Shackleton had expected, but a reprimand for being physically unfit to continue the mission. He felt that there was far more to his release from duty than just his health. Perhaps Scott simply had enough of Shackleton!

It was true that Shackleton's health was questionable since his second bout with scurvy, but it was apparent that there was more to it than that. Scott disliked Shackleton's popularity with the other men. Many of the leadership qualities that Shackleton possessed, Scott lacked, and this threatened his leadership position. One thing was certain, this was Scott's expedition and not Shackleton's, and there could be only one leader! Illness on the southern journey was an excellent excuse to rid himself of Shackleton, and Scott took it. He could also use Shackleton's illness as an excuse for not achieving a more impressive record for Furthest South. Scott could say that it was his concern for Shackleton that prevented him from going further. This was as far from reality as could be, but it made a good story for the rest of the world to read.

What did the Antarctic experience with Scott do for Shackleton? Foremost, he planned to lead his own expedition to Antarctica and capture the South Pole before Scott could do it. By reaching it first, he would show the world how incompetent Scott really was and prove himself to be the better man. Unfortunately, Shackleton was letting his emotions get the best of him. He quickly forgot the lessons that Antarctica had taught him about exploration. For all the errors made by Scott, Shackleton would only improve in his leadership ability. He kept the basic philosophy and approach that Scott had toward exploration. Shackleton too failed to recognize the value of using skis and dogs. Holding fast to this traditional British approach would cost him the immortality of being the first person to stand at the South Pole.

CHAPTER FIVE

HOME FROM THE ICE

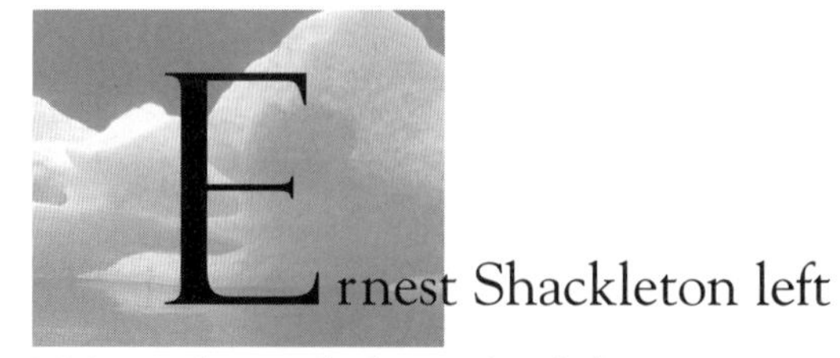

Ernest Shackleton left Antarctica on March 3, 1903 on board the relief ship *Morning*. He was an unwilling passenger. Scott and most of the *Discovery* Expedition crew would remain for another year. Their ship was firmly frozen into the ice and could not escape. To the rest of the world this scene painted a romantic picture of Scott's expedition. It portrayed a trapped ship, a stranded crew and returning seamen that included an invalid lieutenant named Shackleton. In addition to creating an exciting situation for the world to ponder, another year in Antarctica also gave Scott more time to produce some exciting results. That would certainly justify his expedition and make him a public hero. Things were very different for Shackleton. He felt disgraced by his early return and bitterness toward Scott for letting their personal differences get in the way of achieving the expedition's goals. All he could think about was how he could return to Antarctica and make up for the embarrassment he felt.

With Antarctica behind him for the moment, Shackleton continued his recovery in New Zealand. His mind was full of thoughts of both the past and of his dreams for the future. The disappointment of the *Discovery* Expedition certainly did not change Shackleton's feelings for Emily Dorman. It was on board *Discovery* en route to Antarctica that he received word from her father that he had his permission to marry Emily. Since that time, Mr. Dorman had died and now it was solely up to her if they were to marry. He was sure that his misfortune in Antarctica would not matter to her, and he was right. Emily anxiously awaited his return to England.

Homecoming came for Ernest Shackleton on June 12, 1903. When Shackleton arrived in England there was no celebration to greet him as he once imagined. What greeted him was news that a scandal had developed over the affairs of the *Discovery* Expedition. Scott's original plans stated that he was not to remain in Antarctica for a second year. Since he was not able to leave, Scott was now being accused of being an incompetent officer for endangering his ship and crew. In addition, the question of funds to pay for the extra year and a possible relief ship became a major issue. In the absence of Scott, Sir Clements Markham bore the brunt of public criticism. Now he called for Shackleton to provide an explanation for what was happening in Antarctica. What Shackleton had to say was not very pleasing to Sir Clements. Many others were also anxious to hear what Shackleton had to say, including the British Admiralty.

Morning, the ship that carried Shackleton from Antarctica, was preparing to return to provide support for Scott. The Admiralty insisted that a second ship, the *Terra Nova*, accompany her carrying specific orders for Scott. He was ordered that if *Discovery* could not be freed from the ice she was to be abandoned and all crew returned home. It was even suggested that Shackleton should command *Terra Nova* in her relief efforts. Shackleton declined, but did participate in her preparations.

Scott's expedition was not the only one in trouble at this time. Shackleton was also asked for advice and assisted in the preparations for a relief ship to find the *Antarctic* and her crew. She had not been heard from since entering the Weddell Sea. There was great public excitement over these two Antarctic expeditions and Shackleton became an instant expert based on his recent experiences with Scott.

Shackleton enjoyed the limelight and it made up for the embarrassment he felt for being ordered home as an invalid. The thought of being the rescuer of Scott tempted Shackleton to return to Antarctica sooner than he planned. He took great pleasure in the idea of rescuing Scott, the one who proclaimed him unfit for Antarctic service. But now was not the right moment for Shackleton to make amends for his earlier disappointment. There were other matters that had priority, and one was to marry Emily Dorman. His ambition sought fame, fortune and social position and all these could be his if he planned carefully. Antarctica must wait for now.

One benefit Shackleton got from being sent home early was that he could tell his version of the expedition while Scott was still in the Antarctic. Shackleton was already giving public lectures and making a name for himself. Shackleton also received commissions to write articles about his experiences from several newspapers and magazines. Although no direct attacks were made upon Scott's character or quality of leadership, Shackleton was able to convey the shortcomings of both the expedition's leader and those who supported him.

Shackleton also used his celebrity status as a means for meeting wealthy and influential people. He was even able to secure a position as a staff writer with a popular magazine. Journalism was not the road Shackleton wanted to follow to fame and fortune. It is long and hard and this did not suit him. For Shackleton knew that when Scott returned from Antarctica, the spotlight would shift from him to Scott. His day would be over. Something else had to be found in order to build upon his recent success.

Money was one problem that would confront Shackleton all through his career. There was never enough of it to provide him the social status he desired. Although Emily Dorman was reasonably wealthy, Shackleton refused to live off his wife's inheritance. Before they could marry, he wanted to secure a respectable, well-paid position. To this end, Shackleton looked to his friends for help. In England, it was a tradition for a well-to-do gentleman to belong to one or more clubs. Here they could gather for social events, card games and general conversation. Business ventures were often discussed at the club as well as entertaining clients. These clubs grew out of common interests in sports or cultural events. Shackleton sought out his old friends who were interested in adventure or exploration. One such man was Hugh Robert Mill. He encouraged Shackleton to apply for the position of Secretary to the Royal Scottish Geographical Society. Shackleton's appointment to this prestigious society would be the first step on his road to fame and fortune.

By the time he reached his 30th year, Ernest Shackleton had achieved a good measure of success. His desire to marry Emily Dorman was, in part, the reason he had to be successful. He felt that he had to prove himself worthy of her. Antarctica, and all that followed, were just part of his grand plan to win her. Emily, on the other hand, waited a long time for him. She had many chances to marry, but the appeal of Shackleton's ambition and his exciting personality were irresistible. In many ways Shackleton never really grew up. He had that "little boy" quality that Emily found very attractive. She felt that it was her duty to take care of him, for he wouldn't do it for himself. In their own way, Emily and Ernest complemented each other very well. When Shackleton became restless and dreamed of his next expedition, he always knew that Emily would be there waiting for his return. Their marriage took place on April 9, 1904.

The return of Scott from Antarctica signaled the end of Shackleton's importance as the expedition's spokesman. Scott now held the spotlight. There were several occasions when both Scott and Shackleton were present. Although they had made a gentleman's agreement not to show open hostility toward each other, the public could sense the tension between them. Just as there could be only one leader in Antarctica, now there could be only one hero, and that was to be Scott.

Shackleton was not satisfied with the life of an ordinary man, he always needed some new challenge. After Scott's return, he tried to settle into his work at the Geographical Society, but that did not satisfy his needs. Perhaps politics could be the answer for his ambition? In November 1904, he announced that he would seek election to a seat in parliament. This decision created both new friends and enemies alike. Most people thought that he would make an excellent member of parliament and supported his candidacy, while others bitterly opposed his running. Among the opposition were members of the very organization he worked for—the Royal Scottish Geographical Society. They felt that it was beneath the dignity of its members to soil their hands with politics. Even his closest friends in the society thought seeking office was a bad decision. Shackleton eventually resigned his position in the society, and later lost the election too. The fact that he no longer had a job didn't bother him. Shackleton felt certain that influential people would come to his aid, and that his time was yet to come.

Emily gave birth to their first child, a son, on February 2, 1905. Shackleton found adjusting to his new life as a husband and father very difficult. He was restless and dreamed of new adventure. Going back to the same job day after day, year after year just did not appeal to him. He wanted more from life and thought that going into business for himself would be the answer. Shackleton soon learned that he lacked the passion for business and his investments proved no better. He was always looking to make a quick fortune from the money he invested. They all failed, and he was continuously losing money he could not afford to lose. Shackleton was becoming desperate.

At this time in his life, Shackleton appeared to be going nowhere fast. He still placed his hopes in the wealthy and influential people that he had met at the clubs and other social gatherings. Shackleton played the part of a confident man, but he was very unsure of himself. His dreams of a successful business and immense wealth denied him a sense of reality. To make matters worse, the publication of Scott's book, *The Voyage of the Discovery*, placed Shackleton's reputation in serious doubt, and that was all he had going for him.

The Voyage of the Discovery was Scott's official account of the expedition, and it was written totally from his point of view. Readers of the book took every word as absolute fact and it portrayed Scott as a real life hero. Scott's heroic reputation was not earned in Antarctica, but in the publication of his book. References to Shackleton's role in the expedition ridiculed him. Shackleton was especially upset by Scott's references to him being carried along on the sled by Scott and Wilson. There was no mention of the scurvy that affected all of them. Scott also pointed out that it was Shackleton's illness that prevented them from going further south.

Shackleton felt that Scott's book was a deliberate fraud and that it publicly humiliated him. This would be the final act that would make enemies of Scott and Shackleton.

The publication of Scott's book rekindled the dislike Shackleton had for Scott. They were never friends to begin with, but by blaming Shackleton for the expedition's poor showing was just too much. Shackleton was not one to forgive and forget. The only way Shackleton could avenge the damage Scott did to his reputation was to return to Antarctica and better everything Scott had accomplished. Plans began to crystallize in Shackleton's mind. He knew how to organize an expedition and how to lead it, but he didn't know how to raise the necessary money to pay for it.

Shackleton's dreams of returning to Antarctica were flamed by reports of the expeditions of Amundsen and Peary in the far north. These men were following their destiny while he sat in England worrying about money to support his lifestyle. At home, Emily was about to give birth to their second child. Shackleton's interest in his family life had waned and his mind was working feverishly at ways to get back to Antarctica. Reality was not what he wanted to face and family pressures were closing in on him. Escape came from the plans he was making. He believed that all their financial worries would end once he succeeded in Antarctica. Whether he truly believed this or not, it was a comfort to his troubled mind.

Raising money to support a major expedition would not be easy. Scott had great difficulty funding his *Discovery* Expedition, and that was supported by both the government and private contributors. Shackleton would pursue a very different route. He promoted his expedition as a commercial enterprise and looked for potential investors rather than just donors. Shackleton envisioned a series of magazine and newspaper articles, books and even a movie of the expedition. Investors would receive their appropriate share from the revenue that the expedition would generate. Although Shackleton painted a very rosy picture of the financial gains from the expedition, he was haunted by his record of failed businesses and bad investments. He desperately needed a first investor to give his expedition credibility.

William Beardmore was the man who stepped forward and became the first investor in Shackleton's British Antarctic Expedition. Shackleton had impressed Beardmore's wife and she arranged for an interview with her husband. Once Shackleton had the opportunity to explain his plans, it became a relatively easy matter to get Beardmore's support. Now with William Beardmore's support, it became much easier to convince other notable industrialists that his expedition was a worthwhile investment. By early 1907, Shackleton was able to secure the support of many prestigious societies, and even attracted the attention of the King of England. It seemed that the contacts Shackleton had worked so hard to make were finally going to pay off.

CHAPTER SIX

CLOSER
TO
THE POLE

In April 1907, Shackleton came forward and publicly announced his intentions of exploring Antarctica. He was not alone in his quest for the South Pole. Public interest in Antarctica had skyrocketed since Scott's return and there was serious talk of several competing expeditions. The Norwegian explorers, Amundsen and Nansen, were both discussing plans for either a northern or southern polar expedition. A Frenchman, Dr. J. B. Charcot, talked of ballooning to the South Pole, while Henyrk Arctowski acquired Belgian support for his expedition to Antarctica. Shackleton saw Antarctica as becoming a very crowded place and he would have to hurry in order to be the first to stand at the South Pole.

Organizing a polar expedition proved to be more difficult than Shackleton had ever imagined. In selecting a crew, there were few men to choose from who actually had polar experience. Most of the men Shackleton invited were former members of the *Discovery* Expedition. First on his list was Dr. Edward Wilson. They had formed a close bond while serving together on Scott's southern journey, and Shackleton had the highest regard for him. Although Wilson liked Shackleton and wished him success, he declined the offer to participate in the expedition. He felt that it would not be right to come between Scott and Shackleton. Although Wilson considered Shackleton a friend, he was still loyal to Scott despite all his weaknesses. Wilson's decision not to join the expedition was a disappointment to Shackleton, but he continued on with his plans. Other members of Scott's expedition were also contacted. Two did accept, Frank Wild and Ernest Joyce, but all the others chose to follow Wilson's example.

Scott's Discovery *Hut at McMurdo Sound from his 1901 expedition.*

Finding the best crew may have been a problem for Shackleton, but word from Robert Falcon Scott created a more difficult situation. Once Shackleton made his plans public, Scott notified Shackleton that his old base at Hut Point was off limits to his expedition. This angered Shackleton since he had planned on using the *Discovery* Hut as his primary base of operations. This was the logical choice since he was already familiar with the area and could easily move southward from there. Shackleton had envisioned sending out three exploration parties. The first would travel eastward over the Ross Ice Shelf to explore King Edward VII land. A second party would search west toward the mountains of Victoria Land for the South Magnetic Pole. The third party he would lead himself and attempt to reach the South Geographical Pole. Reaching the South Pole was his primary goal and all other aspects of the expedition, including science, were secondary. By making Hut Point unavailable, Scott had severely disrupted Shackleton's plans. Scott even went one step further by declaring his intentions to return to the Ross Sea and claim it as his personal domain for exploration.

Shackleton reacted very badly to Scott's declaration, but there was little he could do about it. After all, Scott was there first, and public opinion was on his side. Although Scott had no legal right to make such demands on Shackleton, professional courtesy forced him to respect Scott's wishes. Shackleton now had to move out of McMurdo Sound and set up his base of operations further east closer to King Edward VII Land. This change caused the Victoria Land component of the expedition to be canceled. It would also make the polar journey both longer and more dangerous, but Shackleton was determined to go through with his plans.

Although changing plans temporarily slowed the expedition's progress, Shackleton was having better luck with finding a good crew. In addition to the two *Discovery* members that already signed on, Shackleton was able to attract several good scientists. Two geologists, an Australian, Professor Edgeworth David, and the Englishman, Raymond Priestly, were anxious to participate in this exciting adventure. Also joining the science team was an Australian physicist, Douglas Mawson. Mawson would later return to Antarctica as the leader of his own expedition. With the core of his science team in place, Shackleton became more optimistic. In spite of all the earlier disappointments and difficulties he endured, Shackleton did have a first-rate expedition and it would sail south.

Nimrod was the name of an old sailing ship that Shackleton purchased for his expedition. She was a small, slow ship that was equipped with both sail and steam engines. Her small size and limited range under steam power created many problems, but Shackleton had to work around these difficulties. It was the best ship he could buy with the money available to him. Several of the earlier pledges to financially support the expedition did not come through and Shackleton had to work within a limited budget.

Choosing the best possible equipment for the expedition was another problem that Shackleton had to face. He did not want to repeat the mistakes that Scott made in his selection of clothing and equipment. Shackleton spent considerable time asking Nansen and other polar explorers for their advice on how to prepare for the expedition. Unfortunately, he was too much like Scott, and did not follow their advice. A good example of this was the use of dogs. Because of the bad experiences they had with dogs on the *Discovery* Expedition, Shackleton chose Manchurian ponies to pull their sledges. Only nine dogs were to be used to support the sledging team that was to go to the South Pole. Also, his reluctance to use skis would be the biggest mistake that would cost him the honor of being first at the South Pole. Time was the main reason that Shackleton decided to go without skis or dogs. Proper dog handling and use of skis requires a great deal of training, and Shackleton did not want to spend the time to learn either one properly. He was in a hurry to get to the South Pole and that's all that mattered.

Shackleton's ship Nimrod *departing Lyttleton Harbor, New Zealand, January 1, 1908.*

Preparations for the British Antarctic Expedition were completed in seven short months. *Nimrod* sailed from the East India Dock in London for New Zealand on July 30, 1907. She was to follow the same basic route as *Discovery* had six years earlier. *Nimrod* was heavily overloaded with supplies and equipment. She even carried an automobile that Shackleton hoped would be the answer to fast Antarctic transportation. He would be very wrong.

The voyage to New Zealand was uneventful and final preparations for Antarctica were made at Lyttelton Harbor, just like Scott. Financial problems continued to haunt the expedition and money was always in short supply. Undaunted, Shackleton looked forward to sailing south so he could be temporarily out of reach from his creditors. His resourcefulness even solved *Nimrod's* limited range while under steam power. He convinced the New Zealanders to provide a ship to tow *Nimrod* all the way out to the Antarctic ice pack. This would save their limited coal supply for when they really needed it in Antarctica.

Before he left England, Shackleton had made a promise to Scott that he would later regret. He agreed not to use Scott's Hut Point base at McMurdo Sound. This would severely limit Shackleton's options once he arrived in Antarctica. As Shackleton sailed past the familiar places he noticed how much change had taken place in just six short years. It became immediately apparent that his plans had to be changed, with the safety of the ship as his biggest concern. Originally Shackleton planned on using the Bay of Whales as his base of operations for his attempt at the South Pole. It appeared to be the shortest and perhaps easiest route. Of course, he could not know for sure until he tried it, but it looked good on the available maps of that day. Once he saw how much the ice barrier had changed, he felt that it was unsafe to spend the winter there. Another, more safe site had to be found, and soon.

Holding to the promise he made to Scott, Shackleton sailed eastward toward King Edward VII Land. All along the way, *Nimrod* encountered large amounts of ice that blocked its path and seriously threatened the ship. With the coal supplies running dangerously low, Shackleton had few options available to him. His first was to turn back and give up his dream. The other was to sail on to McMurdo Sound where it was safe to spend the winter. He knew this would break his word to Scott, but what else could he do? Turning back was out of the question!

Nimrod arrived at McMurdo Sound and found ice blocking its path all the way to Hut Point. Shackleton had to find another location to set up his base. Cape Royds at the foot of Mt. Erebus was the final selection. It was 20 miles from Hut Point, which meant that it would be 40 more miles of travel for the polar journey. In a land where an extra day or even an extra mile could make the difference between

A Gentoo penguin mother keeps watch over her chicks.

Elephant seals meet tourists at Grytvikan, South Georgia.

life and death, the location at Cape Royds definitely worked against Shackleton.

Cape Royds was a rocky site with two nearby small seasonal lakes. It was an inhospitable place to say the least. Landing supplies and equipment was no easy task. An accident involving the hook of a hoist cost Aeneas Mackintosh his right eye. Mackintosh was now on record as having the first medical operation performed in Antarctica. Other problems included horses breaking through the ice, frequent blizzards and the near sinking of the ship. All in all, they had a difficult time establishing their base, but in the end they were prepared to endure the Antarctic winter. There were some pleasant surprises too. The biologists found vegetation clinging to many of the rocks. It was the only life native to the Antarctic continent. Penguins, seals, and birds were all only seasonal visitors.

In Shackleton's plan, *Nimrod* was to depart McMurdo Sound before the ice set in, and remain in New Zealand for the winter. The men at Cape Royds would spend the autumn months much like Scott's expedition did—setting out supply depots. The distance from Cape Royds to the South Pole and back would be 1,730 miles.

Shackleton built this hut at Cape Royds for his 1907–09 expedition. Photo courtesy of Capt. Francis G. Stokes, USN (ret.).

Shackleton intended to cover this distance by having the ponies pull their sledges as they walked alongside. If the need arose, he planned to man haul the sledges all the way to the South Pole. It was definitely his intention to walk all the way. Dogs and skis would not play an important role despite all of the good advice he was given about their usefulness. He was a stubborn man.

Stables for Shackleton's ponies are seen in this view of the hut at Cape Royds. Photo courtesy of Capt. Francis G. Stokes, USN (ret.).

Shackleton used the short time he had before winter set in to test both his men and equipment. His plans depended upon the ponies pulling the majority of the supplies, with his men carrying most of their food with them. There would be no extensive depots to support their effort. All this suddenly changed when four ponies died. This worried Shackleton for he depended upon them to do the job. He did have an automobile available to him, but it was meant to be an experiment in improving Antarctic travel. This was to be the first use of a gas engine vehicle under polar conditions. Although the motor worked fine, the wheels were constantly getting stuck in the snow. The automobile could only be used near base, and even then not very effectively. Shackleton soon realized that he could only depend upon the determination and courage of his men if he was to make it to the South Pole.

The first major accomplishment of the British Antarctic Expedition came in March 1908 when Professor Edgeworth David and his five companions reached the summit of Mt. Erebus. It stands over 12,000 feet high and is covered by ice. Climbing it is a challenge to even the most experienced mountaineers. It took just five days for the party to make its way up to the summit. On March 10 Professor David gazed down into the steam-filled mouth of the active volcano. It was a sight that they would never forget.

Photo of the motorcar that Shackleton brought with him on his 1907–09 Antarctic expedition. From a display at the Canterbury Museum in Christchurch, New Zealand.

Science was always an important part of Shackleton's expedition even though the conquest of the South Pole was his personal goal. To the scientists on the expedition, the discovery of the Earth's Magnetic South Pole was equally important. Unlike the geographical pole which is the fixed point of the Earth's rotation, the magnetic pole shifts with changes in the Earth's magnetic field. Earlier explorers like Ross, Wilkes and Scott failed to find the exact position of the South Magnetic Pole. It required numerous field observations and difficult calculations to find it. Professor David, leading the westward party, went off in search of the magnetic pole toward the high mountains of Victoria Land. Achieving their objective would not be easy. They encountered

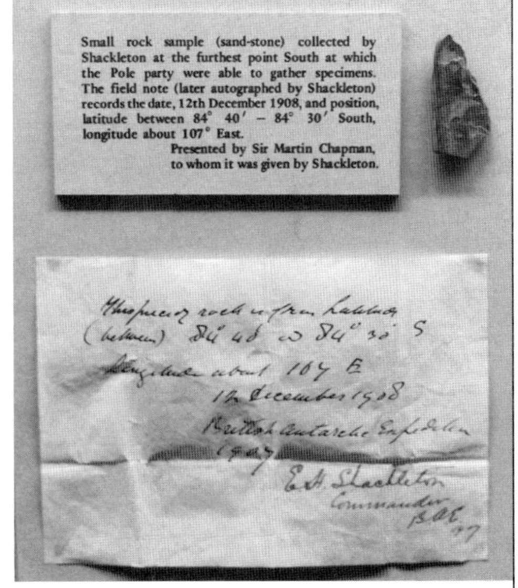

Scientific studies were part of Shackleton's expeditions. This is a rock he brought back from his 1907–09 expedition and is on display at the Canterbury Museum in Christchurch, New Zealand.

many glaciers laced with dangerous crevasses. On one occasion, Professor David fell into a crevasse and had to be rescued by rope. He would not be the only one to fall into these hidden crevasses, and fortunately no one was killed.

The magnetic pole can be located when the dip needle of a compass points exactly downward. Professor David and his men found it at latitude 72 degrees 25 minutes South, longitude 155 degrees 16 minutes East. That is where it was in 1909. Today it is located in the Ross Sea—hundreds of miles from where they originally found it. Their effort was the first major scientific discovery made in the Antarctic. In addition, they made many geological discoveries and laid claim to a vast territory in the name of Great Britain. In doing this, Professor David, Mackay and Mawson completed one of the most successful sledging trips in history. They had covered on foot 1,260 miles of unexplored territory in 122 days. Shackleton was pleased and he felt sure that his expedition would outshine all that Scott had accomplished with his *Discovery* Expedition.

Preparations began in late September 1908 for Shackleton's South Pole attempt. Supply depots had been laid out at strategic points along the route just like Scott had done. Shackleton decided to use Hut Point as his advance base. Supplies and equipment were moved from Cape Royds to Hut Point by both ponies and the automobile. Surprisingly, the automobile outperformed the ponies, but Shackleton never considered using it for the polar journey because it would easily get stuck in the snow. The automobile's time had not yet come for the Antarctic.

Artifacts on exhibit at the Canterbury Museum in Christchurch, New Zealand from Shackleton's Nimrod *Expedition of 1907–09.*

Shackleton's polar journey officially began on October 29, 1908, but they only got as far as Hut Point. Bad weather held them up until November 5th. From there it was on to the South Pole. Accompanying

Shackleton on the journey were Lieutenant Adams, Dr. Marshall and Frank Wild. They carried enough food to last 91 days on full rations. Four ponies would each pull a 600-pound load sledge. For the first 37 miles they were accompanied by a six man support party. They were there to help Shackleton cross a rough section of crevasses. On November 6 the support party turned back leaving the four explorers on their own.

As Shackleton's party slowly crept southward, blizzards and snow blindness made the going very rough. It didn't take long for Shackleton to realize that they were ill-prepared for the hazards that lie ahead. This unexpected change in the weather slowed their progress and forced the men to immediately begin rationing their supplies. If they didn't they would surely run out of food long before they could get back to base. Cutting their daily food rations weakened the men and they were not able to perform effectively. To stay warm in Antarctica a person needs about three times more food than normal to generate sufficient body heat. By cutting their daily food intake, it made them feel cold and weak all the time. Their ponies did not fare well either. Overworked and underfed, one by one the ponies died. Even in death they could still serve an important function. Shackleton marked the position of their bodies for food on the return trip. This he hoped would extend the time they needed to make it to the South Pole and back.

On November 25th, despite the early rough going, Shackleton's party reached 82 degrees 18 minutes 30 seconds South latitude. That broke Scott's record in just 27 days, and it included the extra 20 miles from Cape Royds that Shackleton had to travel. After passing Scott's record for Furthest South, the men saw only

Several glaciers flow out of the Transantarctic Mountains and join the main ice sheet as it flows toward the sea.

new territory. Newly discovered mountains would appear ahead on the horizon and later disappear behind them as they pushed on toward the unknown. Their immediate goal was to find a way through the seemingly endless mountains and hopefully find a flat path to the South Pole. On December 4th they found it. Nestled in a gap between two mountains was a 15-mile-wide, 100-mile-long glacier. It flowed down from a huge ice plateau that lay beyond. Shackleton named this highway of ice for William Beardmore, the first man to offer financial support to the expedition. The maps that Shackleton made would forever record the Beardmore Glacier as the gateway to the South Pole.

The Beardmore Glacier may have been a highway of ice, but it was anything but a smooth one. Progress up the glacier was very slow and difficult. It was full of hidden crevasses that threatened the men at every step. Falling into a crevasse was common and usually not fatal if properly roped together. The ponies were not as lucky. Two died before they started up the glacier, and the last pony met a horrible death as it fell into a crevasse and disappeared from sight. It was by shear luck that Frank Wild was not pulled in too. As it was, they lost half of their food and fuel with that pony. It was a terrible loss that left them doubtful about their chances of reaching the Pole.

A large expanse of darker colored volcanic rock stands out against the lighter colored sedimentary rocks of the Transantarctic Mountains.

Other surprises met the explorers that were not life threatening. One particularly warm day, with temperatures at 22°F, off came the parkas and the men were comfortable enough with just shirts. It felt very warm and Shackleton commented on getting sunburn and frostbite at the same time. Frank Wild also made an interesting discovery. Near the top of the Beardmore Glacier he found an eight-foot-thick seam of coal. Coal always forms in a warm, swampy climate where layers of dead plant material get covered by mud. This was a far different picture of the Antarctica they knew, and it became apparent to them that this continent of ice and snow was once a tropical place.

Christmas Day 1908 found Shackleton and his men inching their way up the Beardmore Glacier. The top was in sight, so they made time to celebrate the holiday with plum pudding, brandy and cigars. Two days later they reached the top of the glacier and looked across a vast featureless surface toward the South Pole. It was 250 miles ahead. If weather conditions were favorable, they could make it. At that moment Shackleton made a bold decision to make a dash south. He now realized that his men were in poor physical condition. They were suffering from bleeding noses, blinding headaches, sick stomachs and suffering the effects of hunger and cold. They were experiencing the effects of altitude sickness—for they were at an elevation of over 10,000 feet above sea level. They were both mentally and physically exhausted.

On January 4, 1909, Shackleton concluded that they could only go on for three more days. Anything longer would risk their chances of getting back alive. He then decided to leave most of their food and equipment behind and go forth with the lightest possible load. By January 6th they were 113 miles from the South Pole at a latitude of 88 degrees 7 minutes South. Here the

On January 9, 1909 Shackleton set a new record for Furthest South. From left to right: Adams, Wild and Shackleton.

weather turned bad once again and their progress was stopped. A blizzard kept them in their tent until 4:00 a.m. the next day. Leaving virtually everything behind they struggled on for another five hours until they could go no further. They had reached latitude 88 degrees 23 minutes South, just 97 miles from the South Pole. After a brief rest they held a short ceremony claiming the Polar Plateau in the name of King Edward VII of Great Britain. Then they left behind a small cylinder containing documents and other mementos from the expedition. It was now time to turn back.

Reaching the South Pole was Shackleton's goal and he did not make it. From where he stood on the Polar Plateau, all he could do was to stare south and wonder. Did he do all that he could to have made it? Should he go on or turn back? Looking ahead as far as the eye could see was a flat wasteland of ice. With great regret he decided that nothing new could be discovered by going on, and it would take all of their efforts just to get back. Shackleton was disappointed, but he would live to tell his tale.

Going back was not going to be easy, but they did have the wind to their advantage. As they ventured southward, the wind was constantly blowing in their faces. Now on the journey back it would be at their backs working with them. In Antarctica the katabatic winds always blow from south to north. With the wind blowing from behind they were able to rig a sail to their sledge and it would easily slide over the ice. This was a big help and they were able to make it back to the Beardmore Glacier in just 12 days. This is where they had depoted most of their food. Going down the glacier would be another matter.

Shackleton and his men were exhausted as they descended the Beardmore Glacier. They were nearly starving and cold beyond belief. In their weakened state they were constantly falling into crevasses and only luck prevented any deaths. Then the worse happened, dysentery broke out. Its symptoms are stomach cramps and continuous diarrhea. They probably caught the disease from eating the raw meat from one of the ponies. Things could not have been any worse.

The weary party now continued its march north, literally surviving from food depot to depot. Fatigue, blowing snow and the possibility of becoming lost all worked against them. When they reached their last depot on February 23rd, they were down to their last biscuits. It was their last hope for survival and they made it. Here they had plenty of food and they ate as much as they could hold, but it was not over yet. Marshall was too weak to continue on and had to be left behind. Shackleton and Wild would go on ahead for help, leaving Adams to stay behind to care for Marshall.

Time was definitely working against Shackleton and his men. Summer was nearly at an end and ice was moving back into McMurdo Sound. By now, *Nimrod* had returned from New Zealand to pick up the men. If Shackleton did not return by the end of February, he was to be considered lost. *Nimrod* could not wait any longer for fear of being frozen in like *Discovery*. These were Shackleton's orders and he knew they had to reach Cape Royds before the ship sailed. Shackleton and Wild had to make a mad dash for the ship if they were to survive.

Back on the ship the situation was somewhat confused. Evans, the *Nimrod's* new captain, knew his orders and he insisted upon leaving as scheduled. Mackay protested their departure and insisted upon sending out a search party for Shackleton. He also wanted *Discovery* Hut to be manned up to the last minute in hopes that Shackleton would show up. All other parties had returned safely and the ship's departure was only a matter of hours away. As Shackleton and Wild made their way to McMurdo, they had no idea if anyone would be waiting for them.

Shackleton and Wild emerged from a blizzard and stumbled into a deserted *Discovery* Hut. There they found a note from Professor David implying that the ship would wait for them only until February 26th, and today was the 28th. Shackleton made frantic attempts to signal Cape Royds, but there was no response and they were too weak and cold to continue. All they could do was eat, rest and accept the fact that the ship had sailed without them, and there were still two men out on the ice.

What was unknown to Shackleton was that a decision had been made to land a winter party at *Discovery* Hut. They would spend the winter there and search for him in the spring to determine his fate. Frank Wild could not believe his eyes as he saw the *Nimrod* coming up McMurdo Sound. Loud cheers could be heard coming from the ship, but Shackleton and Wild were so overwhelmed by the sight that they could not immediately respond. They were going home after all, but not just yet. Marshall and Adams were still out on the ice and had to be rescued. Despite his weakened condition, Shackleton found the strength to lead the rescue party. The next afternoon they reached the depot and brought the two men in.

This dog was accidentally left behind by Shackleton's men as they left Cape Royds and later mummified by the cold dry air of Antarctica. Photo courtesy of Capt. Francis G. Stokes, USN (ret.).

The final days at *Discovery* Hut were spent in preparations for departure. The polar party was recovering well and the scientific results of the expedition were an overwhelming success. Although the Pole had not been reached, they pushed farthest south to within 97 miles of it, and no men were lost. Perhaps better preparations, a closer starting point such as the Bay of Whales, or even better judgement could have given Shackleton the South Pole. But it was not to be this time. *Nimrod* set sail for New Zealand and home on March 4, 1909. The South Pole would have to wait.

CHAPTER SEVEN

YESTERDAY'S HERO

As *Nimrod* sailed north to New Zealand, Shackleton had plenty of time to reflect on his attempt at the South Pole. He felt that the expedition did accomplished its scientific goals, but he was haunted by his failure to reach the South Pole. This was his personal goal. Shackleton had hoped that reaching it would give him the world acclaim that he so desperately needed, and the financial rewards that come with fame. Financial security for himself and his family were very important to him.

Shackleton knew that this time his return to England would be very different from what it was after the *Discovery* Expedition. His attempt was a heroic effort and he did extend the southern record by over 360 miles. The world would praise his accomplishments, but how long would it last? Scott was already planning his return to Antarctica and others were making their plans to explore both the northern and southern polar regions. Shackleton knew that his record would fall, it was only a matter of time. He had to make the most of his popularity while it lasted, and for the moment, let the future take care of itself.

On October 15, 1909 *Nimrod* reached Stewart Island off the southern coast of New Zealand. Finally, after almost two years, they had returned to civilization. Quickly the word of Shackleton's return and the news of his Antarctic journey spread like wildfire. England was proud of Shackleton's accomplishments and it was prepared to give him a hero's welcome. Words of congratulations poured in, notably from Roald Amundsen, Fridtjof Nansen and Otto Nordenskjold. Reluctantly, in public, Robert Falcon Scott acknowledged Shackleton's accomplishments because it was the correct thing to do. Privately, Scott was enraged that Shackleton had broken his word about not using his base at McMurdo Sound. Scott also felt that it was the *Discovery* Expedition that paved the way for Shackleton's success. Even Dr. Edward Wilson, Shackleton's one-time friend and companion on the southern journey, would no longer speak to him. Wilson was also upset about Shackleton breaking his word to Scott. It was a mixed bag of emotions that greeted Shackleton, but there was no doubt who was in the spotlight, and it was Ernest Shackleton.

Just before Shackleton returned to England, word came that Robert Peary and Frederic A. Cook both claimed to have reached the North Pole ahead of the other. Since neither man was available to present their respective claims, Shackleton was still the "man of the hour." Regardless of who actually reached the North Pole first, there would be controversy and considerable debate among the geographical and scientific societies. Shackleton, in the meantime, was the undisputed champion of South polar exploration. His journey was without doubt the story that the public wanted to hear and Shackleton would be quite willing to oblige them. He was a natural showman and took advantage of every opportunity to speak of his adventures.

Aurora Australis *was the book Shackleton wrote after his return from the 1907–09 expedition. Book on display at the Canterbury Museum in Christchurch, New Zealand.*

The public loved Shackleton since he played the part of a real-life hero just as the newspapers and magazines had portrayed him. He had lived a life of adventure in a faraway and exotic place and survived in the face of life-threatening situations. His exploits would fire the imagination of readers everywhere and the public just couldn't get enough of Ernest Shackleton.

Shackleton also became the "darling" of the London social season. He was invited to society dinners and entered into social circles that he had always dreamed about. These were the people Shackleton wanted recognition from, for they could help him with his future endeavors. With his picture constantly in the newspapers, his name appeared at the head of many guest lists. Shackleton was also invited to address the Royal Geographic Society and had audiences with royalty and even the king himself. For the moment, Shackleton's popularity was at its zenith and he loved it.

Of the many honors that were bestowed upon Shackleton, the most sincere and appreciated welcoming came in Norway. He was invited to speak in Christiana at a reception in his honor, and was greeted by a torchlight parade of university students. This was most touching since Norway was the premier country in polar exploration. To receive such praise from one's peers is most gratifying since it comes from those who shared in similar experiences. To have the Norwegians honor him was very special. No one appreciated Shackleton's accomplishments more than Roald Amundsen. In public he acknowledged that Shackleton paved the way for future Antarctic exploration.

Although the public enjoyed Shackleton's vivid descriptions of his expedition, they still recognized that he did not achieve his goal. It was a minor point in all the celebrations, but he did not finish what he started out to do. Not everyone fully appreciated what Shackleton had accomplished, but Friedjhof Nansen certainly did. Nansen held the record for furthest latitude North at 86 degrees, 14 minutes. He could have gone further, but he turned back because weather conditions turned bad endangered the lives his men. Nansen was quoted as saying that "the North Pole was not worth a human life." Amundsen also wanted to be the first to the North Pole, but he too would not sacrifice any life for what he called useless heroism. They recognized the same courage in Ernest Shackleton. With the South Pole seemingly within his reach, Shackleton realized that making it could also mean their deaths. Choosing life over a "hero's" death was a courageous act. Three years later Scott was faced with a similar choice and chose the South Pole and died, taking four others with him.

The public description of Shackleton's journey south did not tell the whole story. Although there was no deliberate cover-up or false claims made, what truly occurred along the polar route remained with the expedition members and was not shared with the public. In reality, some of Shackleton's men were critical of many of the decisions he made on the way to the South Pole. They felt that his obsession to outdistance Scott endangered their lives. Fortunately for everyone, Shackleton did listen to the opinion of others when death stared them in the face.

The expedition actually had several shortcomings that were only obvious to fellow polar explorers. Shackleton's preparations were rushed and inadequate for Antarctic conditions. Much of his equipment was either defective or totally inappropriate for polar conditions. He also chose techniques of polar travel that were clearly outdated and proven inefficient. He walked instead of using skis and dogs. Nothing was more out of place than his use of a motorcar. It was an interesting experiment in polar travel, but its benefit was clearly outweighed by its cost. These were the things that the public never heard about, or even wanted to hear. For the moment Shackleton was the conquering hero that the world cheered, and the truth was not important.

Shackleton's health was the expedition's best kept secret. It was Scott that had used his health as an excuse to send Shackleton home from Antarctica in 1903. It was then attributed to the scurvy contracted on the southern journey. This time, in 1909, Shackleton experienced a similar problem after reaching their farthest point south. He once again collapsed during the march back and had to give up leadership to Dr. Marshall. Shackleton referred to these bouts of illness as being asthma, or a minor heart murmur. It had concerned him for some time, but he refused to undergo any routine physical examinations, and avoided all questions about his health.

Dr. Marshall was aware that Shackleton was experiencing unusual physical discomforts during their march south. He kept a close watch on him long before his collapse on January 21st. After that, Shackleton continued on only by his will to survive. His condition was extremely weak and he suffered terribly. Dr. Marshall had a difficult time diagnosing Shackleton's condition since they were all starving, cold and suffering from dehydration. As in 1903, Shackleton made a remarkably good recovery. He resumed leadership and was able to disguise his recurring health problems. Once they returned to England, Shackleton's health was no longer an issue. Dr. Marshall's concern was dismissed by Shackleton and that was the end of that.

The glory that came from British Antarctic Expedition was soon to fade. No matter who had reached the North Pole first, Peary or Cook, it had been conquered. This drew some of the attention away from Shackleton. When Scott formally announced his return to Antarctica, Shackleton had to give up the spotlight. Once again, contact with Scott could not be avoided. When they would be at the same club or banquet, people sensed the tension between them. Scott was even noted as raising the question of how far south Shackleton actually got. Shackleton could not confirm his actual position since he could not sight the sun. He determined the distance traveled by using dead reckoning and the mileage recorded on his sledge meter. Scott further stated that Shackleton certainly would have made the South Pole if he had been better prepared. These public criticisms were bitter words to Shackleton and fired his resentment toward Scott. Considering all he accomplished, his strength and will to succeed were still being questioned. All this made

Shackleton very restless and moody. Some new enterprise had to be forthcoming, and hopefully soon.

Money was always a big problem for Ernest Shackleton, and creditors were always hunting for him. Upon returning from Antarctica, Shackleton planned on repaying his investors from the money he would receive from lecturing and book writing. But things don't always go as planned. Shackleton soon found himself in deeper debt. *Nimrod* was sold to cover some of the expenses and he went on lecture tours to raise money throughout Europe and the United States. His book, *The Heart of the Antarctic*, was published and some money did come in, but not enough. The financial rewards he dreamed of were not to happen, and he was no better off now than before he went to Antarctica.

Family life for Shackleton was always very difficult. He moved his family to a fashionable address and assumed the role of an accomplished explorer. To most people Shackleton was a success, but he was not happy. Thoughts of beating Scott to the South Pole often drifted through his mind, but for the moment he was tied to his new life. When Shackleton was home for long periods of time he became very restless and uncomfortable. Although he loved Emily and their three children, being at home did not suit him. Shackleton was a much better husband and father when he was away from home than when he was there with them.

Shortly after the birth of their third child in 1911, Shackleton once again suffered severe chest pains. It was probably a heart attack. Perhaps the fast social life he was leading combined with his financial problems finally caught up with him. It was not the happiest time of his life. To make matters worse his reputation was being soiled by the scandals of his brother, Frank. Shackleton had always trusted his brother and had made him the expedition's financial manager. Frank, who was always loyal to Shackleton simply could not stay out of bad investments and questionable business ventures. At one time he was even under suspicion of stealing the Crown Jewels of Ireland. Ernest was constantly borrowing money to keep his brother out of debt, but nothing he could do seemed to help. Eventually Frank was convicted of fraud and was sentenced to 15 months in prison. With so many problems at home it was not surprising that Shackleton preferred the isolation of Antarctica to protect him. Shackleton's only other escape was in the social life of London, and he took special pleasure in the company of several lady friends.

Scott's hut at Cape Evans served as his main base for the 1910–12 expedition. Photo courtesy of Capt. Francis G. Stokes, USN (ret.).

The year 1911 slowly came to an end. Word came that Roald Amundsen had reached the South Pole. The flag of Norway now flew at the South Pole. Amundsen had covered those remaining 97 miles and achieved the glory that had escaped Shackleton. Shackleton eagerly congratulated Amundsen on his success, but stated that this would not be the end of Antarctic exploration. But what of Scott? Scott too had gone to Antarctica with a very well-equipped expedition and had the experience necessary to reach the South Pole. It appeared it was his for the taking. What had happened, and what was Scott's fate?

Once Shackleton had heard that Amundsen was heading for the South Pole, he knew that he would make it. As for Scott, he felt that only luck would get him there first before Amundsen. Shackleton knew that Scott would make the same mistakes he did on the *Discovery* Expedition. The only thing he didn't know was that Scott was trying to beat Shackleton's records, and was not that worried about Amundsen.

Scott planned much of his journey south based on what it would take to beat Shackleton's record. He basically followed Shackleton's route as far as he could and paced himself according to Shackleton's record. Scott was delighted as they passed

Shackleton's mark, but by then it was already too late and they were in trouble. The problem with Scott's plan was that he did not experience the same weather conditions as Shackleton. Scott had overestimated Shackleton's progress. He continued to push on even after he knew that he could not make it to the South Pole in time to beat Amundsen. There was also the question of getting back alive. Scott chose to achieve his measure of immortality in death, rather than what he could achieve in life. Near the end he even altered his records to show that the conditions had beat them and that he did everything right in spite of the results.

The actual South Geographic Pole moves about 30 feet each year due to the movement of the ice. Every year on January 1st the Pole is re-positioned in celebration of the New Year.

Once the news of Scott's death reached England, Shackleton's fate was sealed. Scott was now dead, and right or wrong no longer mattered. Shackleton could not speak out against a dead hero. Scott had finally beaten Shackleton, but only through his own death. It was the book, *Scott's Last Expedition*, that glorified Scott's memory. The book was based on the diary that Scott kept as he was dying. His entries were intended to show how brave they all were and the fact that it was only bad luck that beat them. This is how Scott wanted to be remembered, a fine example of the British character to the bitter end.

The ceremonial South Pole, a favorite photographic stop for visitors and scientists too.

As his dying testament, Scott's diary captured the public's imagination. In it he states that no other expedition had ever experienced

The sign which commemorates the courage of Roald Amundsen and Robert Falcon Scott.

such terrible weather conditions as they did. Nothing could have been further from the truth, but who was there to dispute it? After reading Scott's diary, Shackleton and Amundsen became the real losers. Amundsen would be criticized for making his journey look too easy. Amundsen reached the South Pole first because he used his experience and good judgement to avoid many of the problems faced by Scott. For Shackleton, he would always be remembered for getting close, but not making it all the way. In each case, this was very unfair. Scott, on the other hand, would be immortalized for his suffering and glorious death. It was supposed to reflect the ideal British sense of honor and sacrifice. To Shackleton and Amundsen, Scott's death only proved that he could not achieve the same success in life as he did in death.

Shackleton's life after his successful expedition continued to be a series of ups and downs. First he had to deal with Scott's death and Amundsen's glory. Then, what was to become of Ernest Shackleton? His business ventures were a continuous series of failures, but ironically his financial position improved. The government had agreed to

The main entrance to Amundsen-Scott South Pole Station.

give him a large grant to reduce the debt of the expedition. This was an enormous help. The king also bestowed knighthood upon him, and he became Sir Ernest Shackleton! But his reputation was severely damaged. Scott's book made frequent references to Shackleton's expedition and they were very critical of his leadership and ability. Shackleton never knew that several passages in Scott's journal had been rewritten. Kathleen Scott, Sir Clements Markham and Reginald Smith, the book's publisher, decided to make the appropriate changes that would shift any blame away from Scott. It was now very apparent to Shackleton that his life was going nowhere and the great promise from his Antarctic expedition would not happen.

When things seem darkest, that is when fate often seems to take a hand. Shackleton had no real direction in mind when he received an offer to lead the Imperial Trans-Antarctic Expedition. The idea of crossing the entire Antarctic continent via the South Pole filled the explorer's mind with new hope for glory. This offer also included the promise of a large sum of money to begin his preparations. It seemed like a dream come true. Now as 1913 drew to a close, Shackleton had something very exciting to look forward to in the new year. His star was once again on the rise and he would no longer be just yesterday's hero.

CHAPTER EIGHT

A SECOND CHANCE

Shackleton greeted the year 1914 with great expectations. Ever since he received the promise of financial support from David Lloyd George, Chancellor of the Exchequer, he could not take his mind off the coming expedition. His announcement of the new Antarctic expedition was made public through a letter he had written to *The Times* newspaper. In that article Shackleton stated that a large sum of money had been pledged and plans were moving ahead on schedule. This expedition would not be a mad dash for the South Pole, but a systematic crossing by sledge of the entire Antarctic continent. Nothing like this had ever been thought of before and it clearly displayed Shackleton's ability to catch the public's eye.

The pledge made by Lloyd George was by no means guaranteed since it required Shackleton to raise an equal or greater amount of money. It must also be remembered that it was made by a politician who might not even be in office when Shackleton needed the money. The entire expedition was by no means a certainty, but Shackleton approached it with his usual enthusiasm and salesmanship. It was the challenge that he had been waiting for, and he was not going to let it pass him by.

As news of Shackleton's expedition plans reached the public, it was met with a great deal of excitement. Newspapers all over England began writing about it. Soon the world began to take notice too. Scott's expedition had been portrayed as a noble and heroic effort, but the public soon realized that he came in second and paid for it with the loss of his life and that of his men. Now with Shackleton attempting to cross the entire continent, a new patriotism swept over England. British pride had been bruised by Scott's failure, and many people saw Shackleton as the means for England to recapture its past glory and strengthen its national pride.

Politically, in the early 1900s Great Britain was no longer the unchallenged superpower it once was. Germany's power and influence were on the rise and its navy was rapidly becoming equal to that of Britain. The success of Shackleton's expedition would show the world that the British were still the best. This patriotic fever would also give Shackleton a big advantage in his efforts to raise funds. People who would not normally think of supporting such an expedition would easily give money if they thought it would help the national image. With success, timing is everything, and it appeared that the time was right for Shackleton and his Imperial Trans-Antarctic Expedition.

Although the public was enthusiastic about his expedition, not everyone was supportive of Shackleton. There was an underlying group of individuals who were totally against the expedition. One of his most formidable detractors was none other than Winston Churchill, then First Lord of the Admiralty. Among the most outspoken members of the opposition were Sir Clements Markham and the Secretary of the Admiralty, Evan MacGregor. Fortunately for Shackleton, they were balanced by loyal supporters like Roald Amundsen. Upon hearing news of the expedition, he sent Shackleton a telegram that stated "My warmest wishes for your magnificent undertaking." Endorsements such as this had a much greater positive impact on potential supporters than any of the negative comments.

Raising money was a recurring obstacle to all of Shackleton's Antarctic expeditions. On his first expedition the government made no financial commitments and Shackleton had to rely on private investors. This time Shackleton desperately wanted government support. He already had the promise of a grant from Lloyd George and now he sought support of the British Admiralty. The Royal Navy had been a major financial supporter of Scott's expedition. Scott used his position as an officer in the Royal Navy to his advantage. For Shackleton this would not be the case since he never served in the Royal Navy. What Shackleton asked for and desperately needed was the use of the Royal Navy dockyards. He also requested that the Royal Navy provide him with the necessary stores and equipment to outfit his ship for polar exploration. Shackleton was turned down on both accounts primarily due to Winston Churchill's intervention and general opposition to any further exploration.

Despite all the difficulties Shackleton had in raising money, he did not have that problem finding applicants to join the expedition. Once news of Shackleton's new Antarctic adventure became public, applications flooded the expedition office. Men with a variety of backgrounds and talents applied for the few available positions. Each had their own particular reason for wanting to go to Antarctica. After all, Shackleton was a well-known public hero and people wanted to share in his adventures. Among the first to apply were Frank Wild and George Marston, both members of Shackleton's first expedition.

Emily Shackleton was not at all happy with the idea of Ernest going off to Antarctica again. She had a household to manage and three children to raise. She was tired of doing it alone, especially when money was always a problem. At the start of every expedition, Shackleton had always promised Emily that "this would be my last," but it never was. He enjoyed his time at home with Emily and the children, but only for short periods of time. He just did not fit into family life. Shackleton needed the excitement of an Antarctic expedition. Emily would always give in to his wishes because he was easier to live with when he was involved in an expedition. She was tolerant and supportive of Shackleton and his passion for exploration and was the dutiful wife waiting for his return. She may not have liked this role, but it was the life she had chosen for herself. Surprisingly, it would be those times that she was happiest too.

Shackleton's plan for the Imperial Trans-Antarctic Expedition was a very ambitious one. He would first begin from the Weddell Sea side of Antarctica and then cross the continent by passing over the South Pole. From there it would then be on to the Ross Sea and home. Right from the start the plan seemed flawed for a number of reasons. The plan called for a landing at Vahsel Bay. Then they would immediately set out for the South Pole carrying all the food and supplies they would need to complete the journey. There would be no depots laid out in advance of the sledging party. After their departure, the ship would sail along the Antarctic coast and pick them up on the Ross Sea side. All this had to take place within a single Antarctic summer season. There would be no margin for error. Fortunately for both Shackleton and his men, this plan was quickly dropped.

A more logical plan for the crossing of Antarctica involved the use of two separate ships and sledging parties. One ship would still land Shackleton at a Weddell Sea site, while the other would go directly into the Ross Sea on the other side. It would be the responsibility of the second party to set out supply depots that would support Shackleton from the South Pole to the base at McMurdo Sound. This plan was certainly more sensible than the first, but it still had flaws. The 100 days allowed for the crossing seemed to be too short. To cover over 1,500 miles in such a short time period seemed very optimistic. Amundsen, who averaged 16 miles per day on his spectacular dash to the South Pole, saw Shackleton's plan as unrealistic. It left no margin for bad weather or extremely rough terrain. In addition, the route from the Weddell Sea to the South Pole was still unexplored. Regardless of all the criticism, Shackleton adopted the two-ship plan and trusted in his luck.

The first part of the expedition had to be made by sea, and Shackleton needed two ships that could withstand the rigors of Antarctic waters. Through a bit of good luck, Shackleton heard of a ship in Norway that was specially built for polar waters.

The original buyer had backed out and once it became available, Shackleton stepped in with his offer. It was quickly accepted and he had his first ship. He gave her a new name—*Endurance*. The second ship came by way of his old friend, Douglas Mawson, who sold Shackleton his ship, *Aurora*. Finding a captain and crew for each ship also proved to be quite difficult. Shackleton finally had to settle for Frank Worsley, his second choice, to command *Endurance*, and Aeneas Mackintosh was given command of *Aurora*. Shackleton wanted *Aurora* to be manned by a Royal Navy crew, but his request was denied. The Navy's reason was that war with Germany was almost certain and they would need every available man for the war effort. The Royal Navy did make one concession, they did permit Shackleton to take along seaman who had served with Scott's expedition. It was a token measure of support, and their experience would be useful to Shackleton.

Slowly but steadily the expedition began to take shape. The War Office even permitted several soldiers to join the expedition. They brought with them skills that were definitely needed. Shackleton also received the opportunity to consult military experts on nutrition. He realized that malnutrition and scurvy were the real killers of Scott and his men. Shackleton had to avoid a similar situation. This would be the first time a polar expedition would use vitamins to supplement their expedition food. Another new invention was dehydrated food. It had been tested in Germany and was suggested for the expedition. By using it, they could carry more food with them without increasing the weight they had to pull.

Shackleton's discussions with Amundsen led to another concession—he would learn to ski properly. Amundsen had convinced him that had he used skis in 1908, he would have reached the South Pole. So Shackleton went to Norway and learned to ski. If he didn't use skis for this Antarctic expedition, he would probably die. His trip to Norway was also used to test new equipment such as a motorized sledge and tents that were more efficient. Shackleton was sincerely making every effort to be much better prepared than he was in 1908.

As the date of departure drew nearer, money problems continued to plague Shackleton and affect his preparations. It would be a rich Scotsman, Sir James Caird, that came to his rescue. With a large contribution from Caird, Shackleton satisfied many of his creditors and paid back money that his brother had illegally acquired. With all these money problems haunting him, Shackleton dreamed of the day he would set sail for Antarctica and be free of his creditors.

Much to Shackleton's delight, the departure of *Endurance* would be a public spectacle. A few days before departure, Shackleton received several members of the royal family for a tour of the ship. Despite all the warm wishes and ceremonies that would surround their departure, there was a sense of foreboding hanging over the expedition. It was something that Shackleton had no control over. War clouds were forming over eastern Europe and they threatened to engulf the world. Shackleton's one wish was that he would set sail before war broke out, and for the moment, it seemed that he would get his wish.

Endurance set sail on August 1, 1914. Because of persistent rumors of war being declared, Shackleton anchored off shore and waited for confirmation of permission to leave in time of war. On August 4th, the government ordered the country mobilized in preparation for war. In response to that order, Shackleton sent a telegram to the Admiralty offering his ship and crew for service in the war effort. A quick reply came from Winston Churchill himself; it read "Proceed" and later he confirmed the order in a letter thanking him for his offer. In addition, Shackleton later received a British flag from the king. He had the blessings he was looking for, but not from everyone. Many people felt that the expedition should not sail in time of war, but that would not stop Shackleton. That night war was declared and the world was plunged into four years of horrible carnage. *Endurance* would remain at Plymouth harbor until August 8th when it finally set sail for Antarctica. Little did they know that their world would not be the same.

Although *Endurance* departed England on August 8th, Shackleton did not leave with his ship. He had to temporarily remain in England to further plan the crossing of Antarctica and to deal with persistent financial matters. By now, fighting had broken out in Europe and thousands of lives were being lost on all sides. Shackleton felt some guilt at not being in service with the armed forces, but he was determined to get to Antarctica. In early September, Shackleton saw Mackintosh off to Australia to prepare *Aurora* for her part in the expedition. Shackleton then planned to take a mail ship to Buenos Aires, Argentina where he would join *Endurance*. It was on September 26th that he finally left England, and it would be a long time before he would return.

On October 17, 1908, Shackleton arrived in Buenos Aires and joined the *Endurance*. The ship's voyage from England was rather uneventful, but some crew members were not obeying orders, disloyal, and drunk. These sailors were immediately dismissed and things quickly settled down once Shackleton took charge. His leadership ability was evident as he began to shape his men into an effective crew. They would definitely have to learn to pull together and obey orders if they were to succeed in Antarctica.

Departure from Buenos Aires was a relief for Shackleton for he was leaving behind a world at war and family problems he could not control. He eagerly looked forward to their stop at Grytviken, a Norwegian whaling station in the South Georgia Islands. Here Shackleton would take on fresh water and replenish their needed supplies. Shackleton would also listen to the talk of the whaling captains to learn of the ice conditions around Antarctica. He learned that this season the pack ice had extended farther north than anyone had ever remembered. This was not a good sign. It clearly demonstrated the unpredictability of Antarctica. There was also the very real danger that the ship could become trapped in the ice and not complete its mission. The area Shackleton selected for his operations had previously seen several ships trapped by ice and one ship, the *Antarctic*, sank. He would have to be very cautious and pray that luck was on his side.

CHAPTER NINE

A LONG WAY HOME

It was a reluctant departure for *Endurance* as they sailed out of Grytviken on December 5, 1914. Shackleton appeared to be in no great hurry to leave. Two days after their departure, *Endurance* encountered the ice pack. It came as a big surprise to Shackleton to find pack ice as far north as he did. This would certainly slow down their progress and time was always working against them.

Shackleton's plan was to follow very closely the time table set by Filchner's earlier *Deutschland* Expedition, and he was quickly falling behind schedule. They were proceeding at a rate of one mile per hour through the pack ice. At that rate they would never penetrate the ice far enough south to reach their landing site. The fact that Filchner's ship had been caught in the ice and nearly crushed was also on Shackleton's mind. He had to be very careful that he did not let that happen to him. Unfortunately, luck was not on his side and ice entrapped *Endurance* on January 18, 1915. They were at a latitude of 76 degrees 34 minutes South, and that would be as far as they would get. From that point on, the ship was now at the mercy of the ice and out of control as it slowly drifted northwest away from land.

Endurance *bound by a strong ice floe shortly before being slowly crushed by the surrounding ice.*

Endurance was now a prisoner of the ice. At one point, when they were just 60 miles from their destination at Vahsel Bay, they tried to break free. The crew desperately tried to free the ship by hacking at the ice, but it was no use. They were frozen solid into an ice floe that was three miles long and two and one-half miles wide. There was nothing more they could do but wait and make camp on the ice. This period of relatively calm drifting lasted for three months, then conditions began to change as pressure ridges began to squeeze the ship. The pressure of the ice pushing against the ship's hull began to worry Shackleton. He knew that *Endurance* could not hold forever against the ever increasing force of the ice. If they did not break free very soon, the ship would be crushed and they would be stranded on the ice. The fact that there was nothing they could do about it only made matters worse. They just had to wait and see what would happen next.

On the other side of Antarctica, the crew of the *Aurora* were finally in position at McMurdo Sound. Mackintosh, as leader of the Ross Sea Party as it was called, had sailed from Hobart, Tasmania on Christmas Eve. Due to financial problems in Australia they were over a month late. There was very little money to supply and equip the ship. Mackintosh had to do the best he could and as quickly as possible since he was responsible for depoting the supplies that Shackleton depended on for his return leg from the South Pole. Any delay could cost Shackleton and his men their lives.

Aurora arrived in McMurdo Sound at Cape Crozier on January 9, 1915. Their arrival was not without its problems. As they approached the ice barrier, a fog enveloped the ship and no one could see ahead. At the very last moment the fog lifted and the men saw the barrier directly ahead in their path. A collision was eminent and they did run into it. Quick action on the part of the crew prevented serious damage. Perhaps this unusual incident was a premonition of their future and the part they would play in Shackleton's expedition. Once the world learned of Shackleton's brave journey of survival, the Ross Sea Party would soon become the forgotten part of the expedition.

Landing at McMurdo Sound proved to be quite a difficult task for Mackintosh and his crew. Nine miles away was the closest they could get to their base at Hut Point. To transport their supplies and equipment they would now have to cross dangerous ice. Mackintosh had no choice, and his orders were quite clear. He was to establish supply depots for Shackleton as soon as possible as far south as 80 degrees. After that, they were to spend the winter at both Cape Evans and Hut Point. The following season they were to extend the supply depots as far south as possible. Mackintosh accomplished this first part of his mission, but only with great difficulty and the loss of many dogs.

While Mackintosh was on shore directing operations, Lieutenant Stenhouse was placed in command of *Aurora*. Stenhouse knew that the expedition plan called for *Aurora* to be frozen into the ice for the winter. With this in mind, Stenhouse took his time unloading the supplies and equipment. It would be a costly mistake. Although *Aurora* had been securely moored to the shore, a sudden wind storm tore the ship from its moorings and blew it out to sea. This put the ship in grave danger and it could be smashed to bits by ice floes if they did not act quickly. They had to move out to open water where the ship would be safe. The lifeline to the shore party had been broken and it became impossible to return for the men. The men would be left on their own for the winter and the *Aurora* would have to come back for them next season.

To Mackintosh and his men, the fate of *Aurora* was unknown and they had to carry on with the original plan. They made it through the winter and began setting out additional depots as far South as 80 degrees 30 minutes. It was as far as Mackintosh and his men would get, and their return to base was a very perilous journey. Both Mackintosh and Arnold Spencer-Smith became seriously ill. Because of their weakened condition they had to be left behind. To save everyone's life, three men were sent to bring back food from one of the depots they had just set. They were in serious trouble. A series of blizzards continued to make travel impossible, and the fate of Scott and his party were clearly on their minds. The three men sent back for food did return to their weakened companions, but the situation soon became impossible. In desperation Mackintosh ordered his men to leave him behind and take Spencer-Smith instead. It was not long after that when Spencer-Smith died. The others eventually made it back to Hut Point and, after three days rest, they were able to return and rescue Mackintosh.

Once at Hut Point, Mackintosh and his weakened men ravenously consumed fresh fruit and meat, and they soon began to recover their health. Yet, bad luck continued to haunt the Ross Sea Party. Once he started feeling better, Mackintosh was anxious to rejoin the men at Cape Evans. He insisted upon crossing the soft ice that separated the two parties. This was a very dangerous journey and he was warned against making it. Others in the party refused to go, but Mackintosh and another man were determined to do it. Once they left, concern for their captain prompted the others to follow Mackintosh's tracks. They followed their foot prints for about two miles, and to their horror, they saw the tracks leading straight into a crack in the ice and open water. Apparently Mackintosh and his companion fell through the ice and were lost. They knew that the men could not last long in the icy water and were dead. Things could not have looked worse for the Ross Sea Party. No word had been heard from *Aurora* since being blown out to sea. They had lost their leader and perhaps their ship too. Now they had to settle in for their second Antarctic winter and wait and see what the next season would bring them.

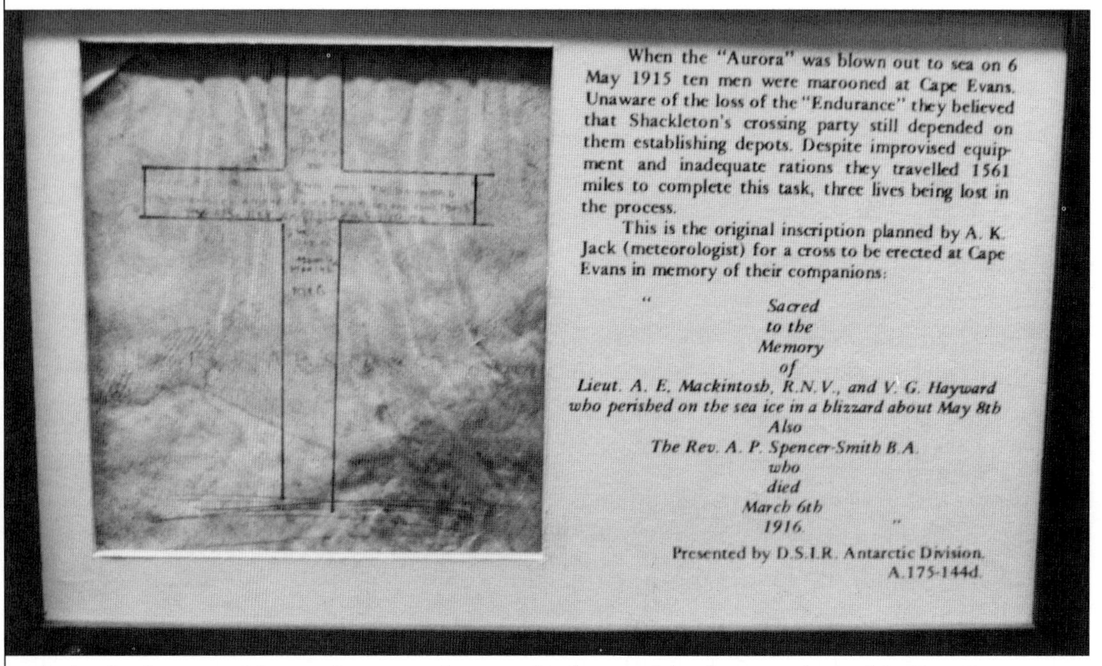

A sketch of a memorial cross that was to be erected in honor of the three members of the Ross Sea party that died. On display at the Canterbury Museum in Christchurch, New Zealand.

Word that the *Aurora* did survive the storm could not be communicated to the shore party, and her fate was still to be determined. Shortly after the storm, *Aurora* became trapped in the pack ice much like *Endurance*. Pressure from the surrounding ice damaged the ship's rudder, but it could be repaired. *Aurora* was still intact, and in March 1916, she broke free of the ice pack and sailed into Port Chalmers, New Zealand for repairs. Now the New Zealand government took charge of the rescue operations. *Aurora* was repaired and rescue operations were finally completed on January 17, 1917. The shore party returned safely after spending over two years in Antarctica. Mackintosh and two of his men were lost. All of their efforts were in vain for Shackleton never reached the supply depots that they placed for them.

On the other side of Antarctica the ice held its tight grip on *Endurance*. For over seven months the ship had been trapped in the ice and had slowly drifted to the northwest. Often Shackleton thought about the progress of the Ross Sea Party. He was confident that if his men had followed his orders everything would go according to plan. Little did he know that their lives were in danger too. But for the present he had to stay focused on his own situation. He knew they had enough food, water, and fuel to survive until they reached open water, and time seemed to be on their side. What concerned him the most was the force and movement of the ice. It made him very nervous. Shackleton could only hope that his ship could withstand the stress and not be crushed by the ice before they could break free into open water.

In the past Shackleton depended upon his skill and some luck to help him through many difficult situations. Now he found himself in a very different situation and it would take more than luck to get him out of it. The relative calm of the slowly drifting ice changed dramatically on August 1st. The ice quickly shifted direction and the floe began to buckle. This exerted a tremendous pressure on the hull of *Endurance* and little by little the ship was giving in to the crushing force of the ice. At first it was noticeable as loud cracking sounds, which later changed to a sharp and sudden cracking noise from shattering wood. Leaks in the hull were now common throughout the ship, and icy water forced the evacuation of many compartments. Around the clock the crew manned the pumps, but the water kept coming in. It was of little use and the ship was slowing being torn apart.

There is some irony in the fact that Shackleton chose the name *Endurance* for his ship. Originally the word "endurance" came from the motto on the Shackleton family coat of arms, "By endurance we conquer." He could never have imagined how appropriate that name would be until his ship had to endure such stress. Although the ship was built to withstand contact with ice, it was never designed to survive the enormous pressure of pack ice continuously pressing against its hull.

The fate of *Endurance* was sealed on October 26, 1915 as Shackleton finally gave the order to abandon ship. It was an emotional decision for Shackleton. Once again his dream of achieving great success in Antarctic was denied, and worse, he also lost his ship. As disappointing as that may have been, Shackleton was faced with a much greater challenge, getting home alive.

Prior to abandoning *Endurance*, the men were living comfortably on board ship and the dogs were camped out on the ice. Now they had to build shelters for themselves on the ice and try to salvage as much food and equipment as they could before the ship sank. Although living on the ice was not as comfortable as the ship, it was not all that bad. They had adequate shelter from the weather, plenty of food, and warm clothing. If the ice remained solid they could survive for a long time. The ice was their main concern and its condition was unpredictable. How long would it last before melting? Would it split open and dump them into the sea while they slept? No one knew the answers to these questions. Shackleton ordered a constant watch be kept and plans were made if an emergency developed.

Shackleton and members of his crew established "Ocean Camp" on an ice floe after abandoning Endurance.

Shackleton quickly developed a plan for their rescue. He knew that they were on their own and that no help would be coming from the outside world, especially since no one knew they were even in trouble. The wireless telegraph, which by now had been in use for many years, could have been useful to Shackleton. He choose not to use it since he expected to be far beyond its useful range. Another fact that worked against help coming to their rescue was that in Antarctic exploration it was common to be out of touch for long periods of time. No one expected to hear from Shackleton for at least a year or more, and he was not yet overdue. The only help Shackleton could count on came from a book. The book was not originally intended to be an Antarctic survival manual, but it turned out to be just that.

Years earlier, Otto Nördenskjöld was in a similar situation as Shackleton when his ship *Antarctic* became ice bound and later sank. Upon his return, Nördenskjöld wrote a book about the sinking and how they managed to survive being stranded in Antarctica. A copy of this book was in the little library on board *Endurance.* It was required reading for anyone interested in Antarctic exploration. Now Shackleton found himself using this book as a source of information on how to survive. During the long wait on the ice floe he read it over and over again, and now it gave him guidance for what to do. The book also mentioned that there was a survival shelter available on Paulet Island some 346 miles from Shackleton's position. It had been used by the crew of *Antarctic* and was stocked with supplies by the relief party that eventually rescued Nördenskjöld. Surprisingly, it was Shackleton himself who had advised the rescue party to leave behind emergency supplies never dreaming that he would be the next one in need of them.

Reaching the shelter was Shackleton's immediate goal. If they could make Paulet Island, he was sure they would be saved. Hopefully the ice floe would continue its drift toward Paulet Island and, as they got near, then they could sledge across the ice to safety. It was a good plan, but it depended upon the cooperation of the

Shackleton and his Endurance *crew spent several months on an ice floe much like these.*

The last look at the wreckage of Endurance *shortly before she sank beneath the ice.*

drifting ice. The ice didn't cooperate. As Shackleton and his men approached Paulet Island, the ice drifted away and kept them from it. Their forward progress was now being canceled out by the reverse drift direction of the ice. Shackleton now made the decision to wait for the ice to break up and then launch their three lifeboats into open water and sail on to safety. They waited and waited, but nothing happened. In despair they named their location Patient Camp. It was an appropriate name.

Endurance lost her battle with the ice on October 27, 1915 as the stress placed against the hull finally crushed it. For 281 days the ship endured and when the end finally came, it was very sad. One could place some blame on the ship's design. She had been built strong enough to resist floating ice, but was not designed to withstand the crushing pressure of being frozen into the pack ice. Amundsen's ship, *Fram*, was a different story. She had a specially shaped hull that permitted her to rise up as the ice pressed in against it. This prevented damage to the hull. If Shackleton had taken better care in choosing a ship, he may have survived the ice. But now was no time for regrets; survival was all that mattered.

Life on the ice floe was difficult. There was ample food from what could be salvaged from the ship, and hunting seals and penguins provided fresh meat. The men did have to be careful not to become a meal themselves. The threat from killer whales and leopard seals was very real and several men were threatened at one time or another. Then there was always the danger of the ice breaking up and the men falling into the icy water. That meant certain death if they were in the water for any length of time, or if the ice closed in and crushed them.

Boredom was another problem that had to be dealt with and activities had to be organized to keep the men active and to occupy their minds. Although the crew of *Endurance* totaled 28 men, life under such close conditions can become quite boring. With little new things to talk about, some men were happy to keep to themselves, while others bonded into small groups. Minor quarrels would break out and Shackleton had to step in to prevent them from becoming serious matters. Shackleton was clearly the visible leader at all times and the men faithfully obeyed his authority. He held the group together with a mixture of discipline and understanding. As the effects of isolation began to take its toll on his men, Shackleton was there to offer hope and encouragement. For himself this was a very difficult role since he had no one to turn to for help about his own doubts and insecurities.

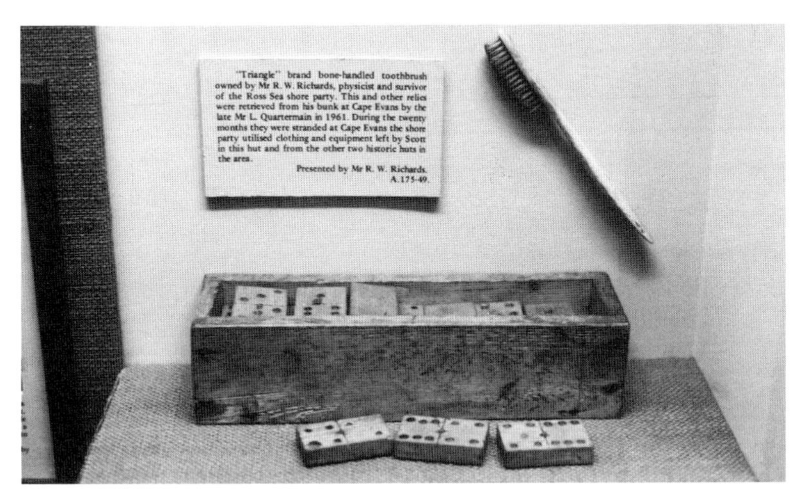

A tooth brush and dominoes from the 1914–16 Endurance *expedition. On display at the Canterbury Museum in Christchurch, New Zealand.*

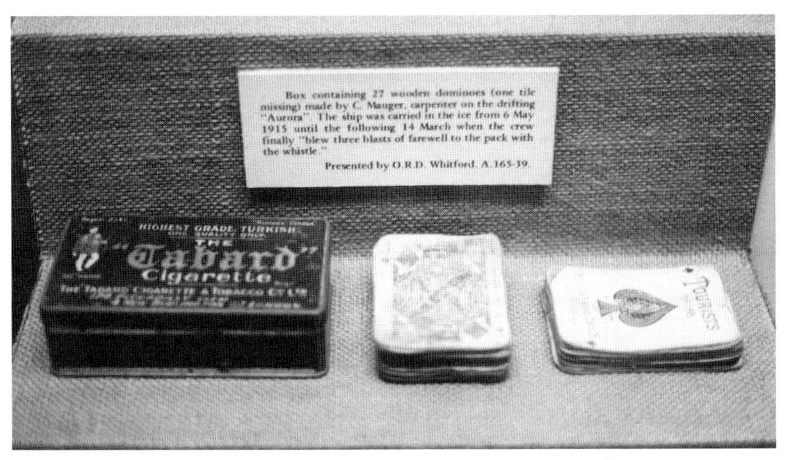

Tobacco and playing cards, necessary items to "pass the time" in Antarctica for Shackleton's men. On display at the Canterbury Museum in Christchurch, New Zealand.

For the crew of *Endurance*, escape from their icy prison came on April 12, 1916. They had been trapped by the ice for

Elephant Island—the first land reached by Shackleton and his men after the Endurance *sank.*

over 15 months when an opportunity to escape presented itself. They were about 100 miles south of two small uninhabited islands. The pack ice was starting to break up and it was now possible to launch the three lifeboats. These boats, the *James Caird*, the *Dudley Docker* and the *Stancomb Wills* were small open craft that could be rigged for sail, but were generally hard to handle. They had been named after some of the more important financial backers of the expedition. Although they were never intended for long journeys, these three small boats were to become the lifeline that Shackleton depended upon to get them home.

The journey from the edge of the pack ice to the nearest island took two days and two nights. Their small boats were designed for use in relatively calm waters and were never meant to sail in the heavy seas of Antarctica. During the journey to Elephant Island, they were hard to handle, constantly took on water, and became covered with ice. Throughout the day and night the men took no rest from bailing water or breaking ice just to keep the boats afloat. There were other problems too. In their rush to launch the boats, no ice was taken on board to melt for drinking water. Since seawater is salty and undrinkable, they were faced with a serious problem. With water all around them, they were all dying of thirst.

After a very rough and difficult voyage to Elephant Island, Shackleton was now faced with the problem of landing. Steering a small boat through rough surf is a difficult task, especially when approaching an unchartered beach. Hidden reefs and shear cliffs could easily wreck the boats. With no safe beaches to make a landing, they desperately searched for what appeared to be their best option. After several failed attempts, the three boats made a safe landing on a small strip of pebble covered beach. The men were exhausted and fell to the ground physically exhausted from what they had just been through. They were now the first men to set foot on Elephant Island since 1830. It was a miracle that all the crew were safe, and now Shackleton could enjoy the simple peace that exhaustion brings with it.

The rugged shoreline of Elephant Island made for a difficult landing for Shackleton, and his men were in constant danger of being washed out to sea even after reaching shore.

Elephant Island was no safe haven, but it was land, the first land they had set foot on since leaving Grytviken on December 15, 1914. Since then they were either at sea or stranded on the ice. That was 16 long months. Throughout that entire period there had been many close calls, but no lives were lost. Now it was time to rest and then proceed with their rescue plans, but first they had to find a safer place to build their survival shelter. The first landing site was too close to the high water mark and it could be easily washed away by storms. A new site had to be found and soon, so once again they had to get into the boats and go back to sea. After some searching a high and dry deserted penguin rookery was found. It was a good place to build their shelter, but it did have one disadvantage, it smelled horrible. It was a small price to pay to be safe on land, and the men settled into making the best of their situation.

Shackleton had two options for rescue available to him. The first had all the men staying at Elephant Island waiting for someone to find them. This plan had many disadvantages since Elephant Island was well off the normal shipping routes and it would be difficult to signal a passing ship. Even if a rescue ship was sent out to find

The launch of the James Caird *on its 700 mile voyage from Elephant Island to South Georgia.*

Shackleton, there was no way of knowing that he went to Elephant Island. By the time a rescue expedition could reach the Weddell Sea, all traces of the expedition would have disappeared with the breakup of the pack ice.

The second plan was even more dangerous, but it had a better chance of success. Shackleton would take five men in the *James Caird* and sail it eastward to the South Georgia Islands. There he would get help from the Stromness Whaling Station and return to Elephant Island for the rest of his men. If he failed to return after a reasonable time, the remaining men had the option to try a second or even a third voyage. Hopefully someone would get through to tell the world what happened. Shackleton chose the second option as the best plan.

The *James Caird* left Elephant Island on April 24, 1916. On board with Shackleton were Worsley, as the navigator, and four of the strongest seamen. They carried six weeks of food and two barrels of fresh water. It would have to last. Everything possible had been done to make the boat seaworthy, but it was still a long shot that they would make it. The boat was cramped and there was little room for the men to move about or sleep comfortably. It would be an unpleasant voyage, but at least they had a fighting chance for survival. Now all they could do was to pray for fair weather and not give up hope.

Back on Elephant Island, Frank Wild was left in command. He instructed the men to turn over both boats and make them into liveable huts. Rocks were later piled around the boats and they were made as weatherproof as possible. These were low cramped quarters, but they did provide shelter against the fierce wind and bone chilling weather. They were trying to make the best of a bad situation.

Among those left on Elephant Island were the two medical doctors. They were constantly treating the many problems that befell the suffering men. The worse medical problem happened to Percy Blackborrow, the youngest member of the crew. He was not a regular member of the expedition, but was actually a stowaway. As *Endurance* left Buenos Aires, Argentina, Blackborrow had hidden below decks and went unnoticed by Shackleton until they were well out to sea. Since then he had served the expedition well, even throughout its darkest days. It was during the boat journey to Elephant Island that his feet became badly frostbitten, and now gangrene set in and threatened his life. There was nothing left to do but amputate the toes on his left foot. If this wasn't done quickly the gangrene would spread throughout his body and he would die. Doctors Greenstreet and Hudson did the best they could with their limited medical supplies and Blackborrow's life was saved. For the moment, all was well on Elephant Island.

An example of the medical kits carried on Shackleton and Scott's expeditions. On display at the Canterbury Museum in Christchurch, New Zealand.

If life on Elephant Island was difficult for Frank Wild and the others, it was far worse for those on the *James Caird*. On the island, boredom was their worst enemy, but the opposite was true for Shackleton and his men. They did not have a moment's rest. Storms beset them from the start and they were constantly being tossed about by rough seas. Several times the ice-

Marooned members of the Endurance *crew left behind on Elephant Island.*

covered boat nearly sank. The rough sea and the ever-changing weather constantly kept the men on their guard. There was no rest and they could not afford to make any mistakes, especially for Frank Worsley. He was their navigator and it was his job to find a very small island in a very large ocean. It was the only point of land where help could be found. The next nearest place was 3,000 miles further east all the way to Africa. They had to find South Georgia, and they had only one chance.

In the past, Shackleton had dreamed of making a long open boat journey and now he welcomed the opportunity. Soon after leaving Elephant Island, his dream turned into a nightmare. As they reached the halfway point to South Georgia, the *James Caird* was hit by a gigantic wave that seemed to come right out of nowhere. It tossed the boat around like a cork and all they could do was to hold on and keep bailing out the water.

Whether it was pure luck or by the grace of God, they survived. They were terrified and exhausted, but they were still alive. Bruised and battered they sailed on. On May 6, 1916, Worsley was able to make the observations necessary to determine their position. They were now less than 100 miles from South Georgia island. After 16 days and nights of fighting the weather and the sea they finally sighted land on May 10th, but it was far from over.

As luck would have it, the path of the *James Caird* had taken it to the uninhabited west side of the island while the help they needed was on the east side. Once again making a landing was no simple matter and, just as before, they had to fight rough surf and dangerous reefs. Once on shore they did not know where they were relative to the whaling station. They now knew that they would have to return to the boat and sail another 150 miles around South Georgia to find the station. Shackleton was not convinced that his men or the boat could make that journey. They were both beat-up, and there was another option. Shackleton could attempt to reach the other side by crossing the high mountains that divided the island. It was certainly a shorter but equally dangerous trip. Neither was very safe, but there were no other choices. A decision had to be made. After much thought Shackleton decided to take the overland route. It would be a difficult 29-mile journey for any climber and nearly impossible for men in their weakened condition. The mountains between them and the station were high and rugged and had never been mapped or crossed before. With little or no equipment they would attempt a climb that would test the most experienced Alpine mountaineer. Their chances of success were unknown, but they had to give it a try.

Three of Shackleton's men were too weak to attempt the trip and had to be left behind. A strong hut was built and there was plenty of food and fresh water available for them. They would be safe enough until Shackleton returned, or until they were strong enough to make their own attempt. The trip over the mountains would be made by Shackleton, Worsley and Crean. With a brilliant moon lighting their way, the trio left at 3:00 a.m. on May 19th. They were traveling light and carried only what was necessary to stay alive. They had a 50-foot length of rope, a make-shift ice axe, and had driven screws through their boots for better traction when crossing ice. That was the total of their climbing equipment. In addition, they had a three day supply of food and a primus stove to melt ice. They had to move fast or die.

For Shackleton, climbing these rugged mountains was as difficult as it had been going up the Beardmore Glacier back in 1908. Had it not been for their will to survive they surely would have perished. They knew that help was just ahead and it kept them going. Cold, tired and hungry, they inched their way up to the crest of the mountain range. It took three attempts before they were able to find a way down to the other side. Time was once again working against them and they would have to throw caution to the wind if they were to survive. Shackleton then suggested that they sit on their coiled rope and use it as a sled and try sliding down the icy slope. It was a bold move that could have easily ended in disaster, but the alternative was to die high up in the mountains. Surprisingly it

The high mountains on South Georgia that Shackleton had to cross to get help for his men stranded on the west side of the island.

The Stromness whaling station at Grytvikan, South Georgia Island where Shackleton sought help for his men.

worked very well. They slid down the slope picking up speed as they got further and further along until they finally leveled off in a valley 2,000 feet below the place where they had started. The gamble worked this time.

The route to the whaling station lay through a vast crevasse field. Any one of these crevasses could bring certain death to Shackleton and his men. Slowly they made their way safely through. Sunrise that morning found the three climbers exhausted and wondering just how much further they had to go and what other dangers awaited them. Suddenly through the crisp mountain air came the sharp sound of a steam whistle. It was the call of the whaling station waking up its men for work. No sound could have been more pleasing to Shackleton and his weary companions. By 1:30 p.m. they sighted the Stromness Whaling Station and it was just 2,500 feet below them. The end was in sight, but the last leg of their journey would be as dangerous as the first. The final descent to the station could only be made by going through an icy stream and over a waterfall. Finally their goal was in reach and the station they had left so long ago was now just a mile ahead.

The foreman of the station was one of the first men to sight Shackleton's party. As he was making his morning rounds of the station he noticed three men walking in from the west. That was unusual since all visitors came in from the shoreline to the east. In addition, these men were a frightful sight having scared two young boys by their appearance. They had long dirty hair and beards, had blackened faces, and wore torn clothes. Their eyes were reddened and sunk deep into their faces. By all accounts they had all the features of a monster. No one could have guessed it was Shackleton. His ship had been given up for lost a long time ago. No one could imagine who these mysterious men were.

The manager's house at the Stromness whaling station Grytvikan, South Georgia.

Gathering up all his courage the foreman approached the men and heard one of them politely ask in English to see the station manager. Cautiously they were taken to his home and when the manager answered the door he stood there in disbelief at what he saw. It was only after some time that he asked "Who are you?" A reply came from one of the men, "My name is Shackleton." To the manager this seemed impossible; this could not be the man he met back in 1914. Once the shock had worn off and Shackleton and his men were given a bath, they were treated to the most wonderful food they ever had. Nothing was too good for these men who had returned from the dead.

An immediate rescue of the three men left on the west side of the island was made and they soon joined the others in the warm welcome of the whaling station. It was a great moment for Shackleton, but his mind was still elsewhere. His thoughts were on the men left behind on Elephant Island. There would be no rest for him until they were safe too. He immediately sent out an urgent request for help but found out that the world of 1916 was very different from what he left behind only two years ago. The world was at war and it thought that he had perished long ago. It was with great surprise that it welcomed him back into civilization.

The station manager's bath tub where Shackleton had his first bath in almost two years.

An answer to Shackleton's call for help was slow in coming so he hired the first whaler he could find and set out for Elephant Island. This ship, the *Southern Sky* was a steel hull vessel and was not built to sail into the pack ice. Seventy miles from Elephant Island they encountered ice and could go no further and turned back. Shackleton looked ahead helpless toward the direction of Elephant Island and wondered if his men were still alive. He could only hope that they could hold on until he could rescue them. Shackleton vowed he would not give up trying until they were all safe.

Shackleton then looked to the British Admiralty for help, but sadly it was denied. They insisted that all available ships and men were involved in the war and nothing could be spared. Shackleton then found help elsewhere, but two more attempts also failed to reach Elephant Island. It seemed that things were looking very bad for the 22 men left behind, but Shackleton would not give up. He finally convinced the Chilean government to lend him the ship *Yelcho* to make yet another rescue attempt. This time luck was on his side and they approached the island with little difficulty. The date was August 30, 1916. On deck an excited Shackleton peered through his binoculars searching the shoreline for signs of life. One by one he counted 22 men. Unbelievably, all hands were safe.

Rescue operations began at once. The men on shore had been there for almost 20 weeks and were near starvation. They had no idea that Shackleton had even made it to South Georgia Island. Despite their weakened condition, the men had thought about trying their own escape, but their real

Cheers rang out from Shackleton's marooned men as the rescue boat approached Elephant Island on August 30, 1916.

hope was placed in Shackleton's return. Great joy overcame the men as they sighted the rescue ship. "The boss" had returned just as he promised, and only just a little late. Finally, all the men returned to civilization. Haircuts and baths were the order of the day. Later they were greeted by the sounds of brass bands and cheering crowds. People just couldn't do enough for them. They were heroes even though the Imperial Trans-Antarctic Expedition had been a disaster.

Shackleton enjoyed the celebrations that he and his men were given, but the big question still on his mind was the fate of the Ross Sea Party. He knew that *Aurora* had been blown out to sea and later trapped by ice just like *Endurance* had, but she managed to escape. The fate of Mackintosh and the shore party was another matter. Nothing was known about them and Shackleton had to find out for himself. He reached New Zealand just before *Aurora* left on her rescue mission. For Shackleton, this time it would be a sad return to McMurdo Sound. His great plan for the crossing of Antarctica had failed and it was only through luck and determination that he got his men back home alive. Things were not so lucky for the stranded members of the Ross Sea Party.

The sight of the men at McMurdo Sound was even worse than that of the men on Elephant Island. Three men had died and those who survived were in very bad shape. Even though they had plenty of food and water, isolation and lack of proper clothing had taken its toll. These men spent over two years in Antarctica wondering if their rescue would ever come. They felt abandoned and many blamed Shackleton for their situation. Upon seeing his men Shackleton was saddened and accepted responsibility for all that happened. He fully realized that all their suffering was for absolutely nothing. Shackleton was anxious to leave McMurdo Sound behind him with all its memories of disappointment and failure. As *Aurora* sailed north for the last time the thought of what the future held for Ernest Shackleton was not very important to him. What was important was the fact that all his men were finally safe and on their way home. He could now return to England and pick up the pieces and get on with his life.

CHAPTER TEN

THE LAST VOYAGE

THE LAST VOYAGE

Shackleton's return home from Antarctica was not what he had envisioned back in 1914. World War I held the attention of most of the world and the Imperial Trans-Antarctic Expedition, after all, was a failure. News of Shackleton's heroic voyage and the rescue of his men made world headlines, but they quickly faded in light of the bloody battles being fought throughout Europe. The 53 survivors of Shackleton's Antarctic expedition all entered national service. Many of the men joined the war effort believing it was their patriotic duty to serve their king and country. Others joined to try and forget their ordeal and make up for those lost years on the ice. The war took its toll on Shackleton's men killing three and wounding five in battle.

Although Shackleton's expedition did not accomplish its goal of crossing Antarctica, it did have many positive results. Numerous scientific observations were made while they drifted on the ice floe. They measured ocean depths, recorded weather conditions and added 300 miles of new coastline to the map of Antarctica. Their experience on the pack ice provided a great deal of information about its drift and the mechanics of ice movement. The expedition certainly was not a total loss and it was remarkable that they acquired any data at all.

This time around Shackleton's return to civilization was different in many ways. For the first time he was not besieged by creditors. Thanks to the efforts of Leonard Tripp, the expedition's business manager, Shackleton did not have to face a mountain of debt like he did in 1909. The expedition's financial matters were well in hand and all the debts would be paid. Unfortunately that was not the case with his personal life. Debt was still his constant companion and he owed money to both family and friends. It was an all too familiar situation for Shackleton and he resented his inability to pay the debts. Although he had several plans to raise the needed money, there was no guarantee that he would succeed. Shackleton decided that his best chance to raise fast money would come from writing a book about the expedition and going on a lecture tour in the United States. At first the lectures went very well, but public interest quickly declined after the United States entered the war in 1917. As for the book, Shackleton hated the monotony of writing so he hired a ghost writer to make something out of his notes. Things were simply not going the way Shackleton had planned, and clearly his life was no better now than before he left for Antarctica.

Family life too proved to be very difficult for Shackleton. The normal day to day routine of family life was boring for him and he became increasingly restless. Although Emily tried her best to make a happy home for him, it was no use and he slowly drifted away from the family. Shackleton spent a great deal of time at various clubs in London and he made the social scene his priority. Both his mind and body seemed to be failing him as his depression grew. Heavy smoking and drinking finally began to take its toll on his body and he was often very ill. A life without the dreams of glorious expeditions or exciting financial schemes made Shackleton a very unhappy man. He had to resort to living by his wits and depending upon the charity of others. This was certainly not the man who brought 28 men safely back from the wilderness of Antarctica.

Despite his depressed condition Shackleton was still a prominent, well-respected public figure. He also had many good friends in government who wanted to help him. When an offer came for him to enter diplomatic service, he jumped at the chance. This opportunity would eventually take him to Russia and Argentina. Later he became involved in a business enterprise known as the Northern Exploration Company. It was an invention of the British government to give it a presence on Spitsbergen, a strategic island north of Norway. Shackleton was the perfect man for the job and he welcomed the opportunity to cross the Arctic Circle. It was now obvious that Shackleton was happiest when he was farthest from home.

It was on the trip to Spitsbergen that Shackleton became seriously ill. On several other occasions in both Antarctica and at home he had taken ill from some unknown cause. Each time he had always refused any treatment from doctors and would not permit them to examine him. Still, to Dr. McElroy, his friend and companion from *Endurance*, his condition strongly suggested heart disease. Shackleton would not hear of it and resumed his normal lifestyle once he was sufficiently recovered from his latest attack. He was a very stubborn man when it came to questions about his health.

World War I finally came to an end in November 1918. It had changed the face of Europe forever and in the process had affected the lives of millions of people. Shackleton did his part and had served his country well during the last few months of the war. He had made many contacts during his service and there was always hope of possible business ventures, but none materialized. It seemed that Shackleton just could not get that one break that would assure his financial independence. At war's end he was a very disillusioned man. From December 1919 to May 1920, Shackleton resorted to appearing on stage twice a day giving lectures based on his experiences from *Endurance*. Often he would be speaking to only a few people for very little money. This was the lowest point he would reach in his career and at the same time, the exploits of Scott were becoming legend.

South, Shackleton's promised book describing the voyage of *Endurance* was finally in print. Shackleton, after dictating its text, actually had very little to do with writing the book. The writing was done by Edward Saunders and edited by Leonard Hussey. The book sold well and it was successful, but Shackleton realized none of the money. His share of the profits had been promised to his creditors. Lack of money continued to be Shackleton's most pressing problem and he continued to become involved in bad business ventures. Nothing seemed to be going his way. His marriage to Emily was in name only and he would only spend time with his family when it was convenient for him. Shackleton's health also took a turn for the worse. He was constantly suffering from colds and fever. The two years of physical and mental stress from his *Endurance* experience could not compare to what his present lifestyle was doing to his physical condition. His overeating, drinking and smoking were clearly catching up to him and it showed in his appearance. To his old friends from the Antarctic days, it was hard for them to believe this was the man who had saved their lives. Perhaps now someone had to step in and save him from himself.

By the spring of 1920, Shackleton seemed to have lost all sense of direction in his life. He still enjoyed the respect he received in social circles, but it was becoming an empty presence. This was not the same world that once hungered for news of Antarctic exploration. No longer were Shackleton's Antarctic exploits front page news, and he felt that he was living on yesterday's glory. He hoped that success in business would bring him the satisfaction he so desperately needed. An attractive business opportunity in the Canadian Arctic caught his attention, but it was developing too slowly to hold his interest. When all of his other business ventures failed Shackleton once again looked south in hopes of recapturing his former glory. He was now in a big rush to get as far away from England as possible and Antarctica was waiting.

It would be his old school friend, John Quiller Rowett, who would come to Shackleton's aid. Rowett had become very rich in the rum trade and he wanted to get involved in something more exciting. It was easy for Shackleton to convince Rowett to finance an Antarctic expedition. They agreed to share equally in all the rewards the expedition would bring them. The main goal of the expedition would be to circumnavigate the Antarctic continent. In doing so, Shackleton hoped to discover new islands and to explore a poorly-known section of the Antarctic coastline. In addition, he proposed looking for buried treasure, searching for pearls and finally studying the native cultures of the Pacific Islands. In other words, Shackleton was looking for any excuse to go south again.

Quest was the name of the ship that Shackleton chose for the expedition. It was small and poorly designed for long sea cruises, especially in polar waters. By all standards it was inadequate for Shackleton's purpose. For a crew, Shackleton called upon many of the men from *Endurance*. Most of these men expressed no interest in going on the expedition, but they accepted because of their loyalty to Shackleton. They would follow him anywhere he asked, because to them, he was still "the boss."

The *Quest* Expedition set sail on September 17, 1921. A large crowd was present to cheer them on as she sailed down the Thames River in London. Emily Shackleton was not there this time to say goodbye. As usual, Shackleton did not leave Emily enough money to support the children or run the household. To make ends meet she had to depend upon the generosity of Janet Stancomb-Wills, one of Shackleton's financial supporters. Even Rowett, who funded the expedition, did it because he pitied Shackleton. Shackleton had asked him for the money because he had no one else to turn to for support. Had Shackleton known this, it would have seriously hurt his pride.

Once the *Quest* left port, things did not get any better for Shackleton. All was not well on the voyage from England to Rio de Janeiro, Brazil. Under heavy seas the ship sailed badly, and it soon became apparent that the expedition was poorly organized. Shackleton also became quite ill for most of the voyage. In spite of his condition, he would still not permit a doctor to examine him, not even his friend Macklin from the *Endurance*. Upon reaching Rio de Janeiro, Shackleton suffered a massive heart attack. It was now apparent that his condition was serious, yet he would not call off the expedition. Shackleton insisted that they continue on to Grytviken in the South Georgia Islands. This was his old embarkation point for Antarctica and he was anxious to get back. During the journey to Grytviken, Shackleton ignored his condition and did not slow down in his responsibilities to the ship. During a particularly bad storm he remained on the bridge for four consecutive nights. He felt it was his duty to see the ship through this danger all by himself.

Quest arrived at Grytviken on January 4, 1922. Ironically this was the exact date that the *Endurance* arrived eight years earlier. There were still all the old familiar buildings and a few old faces to greet him. All the old memories quickly came back to him and he felt at home. It would be just like old times and the fire of exploration once again burned brightly in his mind. It was with great joy that Shackleton stepped ashore and met with station manager, Fridtjof Jackobson. Together they remembered how it was that day in 1916 when Shackleton first appeared at his door after his incredible voyage. Here at Grytviken, among the old sailors of the southern seas, he was still the Shackleton of old and not the worn out old man that he had become.

The end of life's long voyage came for Ernest Shackleton in the early morning hours of January 5, 1922. He had suffered another heart attack and this time he did not recover. He spoke his last words to Macklin and died. Perhaps it was fitting that he died in a place he loved so well and among the men who knew him as the man he truly was. When word of Shackleton's death reached Emily she gave directions that the men bury him near the Stromness Whaling Station on Grytviken. The site they chose was close to where he emerged from the mountains back in 1916. They all felt that Shackleton would have liked that. Without "the boss" the *Quest* Expedition was over and returned to England. With the death of Ernest Shackleton the "Golden Age" of Antarctic exploration came to an end. Soon airplanes would fly over the ice and technology would make the expeditions a little easier but no less dangerous. Each person who followed in the footsteps of Shackleton, Amundsen and Scott would remember that they were there first.

EPILOGUE

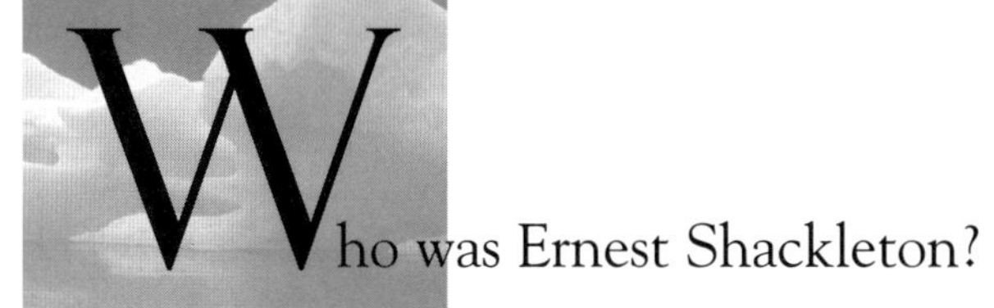

Who was Ernest Shackleton? He was an ambitious man who had great dreams. It is true that he was not a success in the strict sense of the word. His three expeditions to Antarctica all failed to achieve their goals. So why do people remember him as a hero? Perhaps the reason is that he tried his best, but just fell short of success. It seems that he was destined to show the way that others would follow, but to never reach that goal himself. Shackleton's farthest south in 1909 paved the way for Amundsen who reached the South Pole in 1911. Shackleton's plan for crossing the Antarctic continent would finally become reality in the austral summer of 1957–1958 when Sir Vivian Fuchs and Sir Edmund Hillary completed that journey. The example set by Shackleton not only inspired future generations of Antarctic explorers but has carried over into the modern business world. Business executives look to Shackleton's *Endurance* experience as a prime example of crisis management and his style is taught to their employees. The wisdom of the old explorer lives on.

Although Ernest Shackleton may have made serious mistakes in the way he organized and prepared for his expeditions, his true worth came in the face of adversity. His leadership ability was by far his best attribute. His destiny was to lead and not to follow. This was evident on Scott's *Discovery* Expedition. Those men turned to Shackleton rather than Scott as their natural leader because he displayed the ability to make the right decision when danger threatened. The men who followed Shackleton knew that their welfare and safety were first in Shackleton's mind, and that's why they trusted him. Shackleton may not have always been right, but he never blamed anyone else for his mistakes. He always took full responsibility and did not make excuses. This quality set him far apart from him and men like Scott.

A copy of a portrait of Shackleton made by George Marston who served as artist for the Endurance *expedition.*

Even though some of Shackleton's men may have looked upon him in a god-like manner, he was anything but a god. He was just a man like any other and he had his faults. Although he loved his wife and family, he was not the best husband or father. His restlessness and ambition would not let him play the role of a traditional family man. Chasing rainbows was more to his liking.

History may have judged Ernest Shackleton rather harshly because of his record of failures. Those who knew him as "the boss" or served with him on other expeditions knew better. Amundsen may have been the better expedition organizer, but it would be Shackleton who would be the better leader if survival were at stake. Shackleton did not know the meaning of the word quit and he was always there to urge his men on just when everything seemed hopeless. For that he is best remembered and respected by the people who served with him, and those who followed in his footsteps. As each Antarctic expedition stops in at Grytviken in the South Georgia Islands, they place a wreath or medal on the grave of Sir Ernest Shackleton as a sign of their respect. He is not forgotten!

The grave of Ernest Henry Schackleton at Grytvikan, South Georgia. Expeditions passing through on their way to Antarctica leave mementos behind in honor of "The Boss."

BIBLIOGRAPHY

BIBLIOGRAPHY

Alexander, Caroline. *The Endurance-Shackleton's Legendary Antarctic Expedition.* Alfred A. Knoff, New York, 1998.

Bickel, Lennard. *Shackleton's Forgotten Argonauts*. MacMillan, 1982.

Bickel, Lennard. *Shackleton's Forgotten* Men. Thunder's Mouth Press & Balliett & Fitzgerald Inc., New York, 2001.

Chapman, Walker. *The Loneliest Continent*. Graphic Society Publishers, Ltd., New York, 1964.

Cherry-Garrard, Apsley. *The Worst Journey in the World.* Chatto and Windus, London, 1965.

Honnywell, Eleanor. *The Challenge of Antarctica.* Nelson, Oswestry, Shropshire, England, 1984.

Huntford, Roland. *Shackleton.* Fawcett Columbine, New York, 1985.

Huntford, Roland. *The Last Place on Earth.* Athenaum, New York, 1986.

King, Peter. *South: The Story of Shackleton's Last Expedition, 1914–17: Sir Ernest Shackleton*. Century, Ltd., London, 1991.

Lansing, Alfred. *Endurance—Shackleton's Incredible Voyage.* Carroll and Graf Publishers, Inc., New York, 1986.

Lansing, Alfred. *Shackleton's Valiant Voyage.* University of London Press, London, 1963.

Morell, Margot and Stephanie Capparell. *Shackleton's Way*. Viking, New York, 2001.

Newman, Stanley. *Shackleton's Lieutenant: The Nimrod Diary of A.L.A. Mackintosh British Antarctic Expedition 1907–09*. Polar Publications, Auckland, 1990.

Shackleton, Ernest. *Aurora Australis*. SeTo Publishing Ltd., Auckland, reprinted 1988.

Sipiera, Paul P. *Roald Amundsen and Robert Scott—Race for the South Pole*. Childrens Press, Danbury, 1990.

Thompson, John Bell. *Shackleton's Captain: A Biography of Frank Worsley*. Hazard Press, 1998.